the Solar
HYDROGEN
civilization

by Roy McAlister

T his book is dedicated to the kindred spirit of progress provided by every person since the beginning of human time, and every person who will exist until stars no longer support life. During this brief time in cosmic events in which we live, humans will surely record a remarkable evolution from dominance by the greediest to prosperity by the enlightened.

Let that time be now and the achievement be yours.

Regarding near-term issues, this book is to implore action by every member of the global community who wants to achieve security for each and every homestead, nation, continent, and the world.

Sincere participation in The Solar Hydrogen Civilization will facilitate worldwide prosperity and overcome threats of malnutrition, disease, conflict, catastrophic climate change, and mass destruction by collision with objects in space.

Grand Plan for Achieving Sustainable Prosperity

Graduates from the world's leading universities, colleges, and trade schools along with qualified veterans who earn honorable discharges from military service are solicited to join the best, brightest, and most determined farmers and other entrepreneurs to form 6,500 new ventures and attract investment for providing key technologies and services to facilitate sustainable economic development in communities of approximately one million persons.

The Grand Plan, unifying purpose, and destiny of humans on Earth is to provide as much hard work, inspiration, and dedication as it takes to achieve sustainable prosperity in every community by harnessing solar, wind, wave, falling water and biomass resources to overcome Civilization's futile dependence upon burning over one million years' of fossil accumulations each year.

Contents

Illustrations by Michael Swaine

Published by the American Hydrogen Association
P.O.Box 41896, Mesa, AZ 85274
www.clean-air.org
www.americanhydrogenassociation.org -or- www.GoH2.org

Second Edition
ISBN 0-9728375-0-7
Copyright © 2005
Roy McAlister

Acknowledgements

Our standard of living is better, our hopes more realizable, and the miracles of the Earth, Stars, and the Universe are much more spectacular because of their insights, inventions, inspirations, and discoveries.

Irst, I would like to thank the following persons and organizations for their outstanding pioneering efforts and contributions to the advancement of the Solar Hydrogen Civilization.

Theophrastus Paracelsus	H_2 from metal/acid reactions	≈1530
Galileo Galilei	Laws of Motion, Telescopes	1604
Isaac Newton	F = M(a) and Calculus	1666
Henry Cavendish	Identification of Hydrogen	1776
Jan Ingenhousz	Photosynthesis Analysis	1778
Jacques Charles	27-Mile H_2 Balloon Flight	1783
Antoine Laurent Lavoisier	Named Hydrogen	1785
Jean-Pierre Blanchard and John Jeffries	Crossed English Channel by H_2 Balloon	1785
William Murdock	H_2 Production, Pipeline Delivery, Gas Lamp	1792
Anastasio Volta	Electric Battery	1800
William Nicholson and Sir Anthony Carlisle	Electrolysis of Water	1800
Francois Isaac de Rivaz	Designed First Hydrogen Engine and Car	1807
Robert Stirling	Hot Air Engine Economizer	1816
Sadi Carnot	Carnot Cycle: $E=1-(T_L/T_H)$	1822
Michael Faraday	Laws of Electrolysis and Circuits	1833
Sir William Grove	Reversible Electrolyzer/Fuel Cell	1839
Herman von Helmholtz	Law of Energy Conservation	1847
Jean Joseph Lenoir	Invented First Practical H_2 Engine	1859
Thaddeus Sobieski Lowe	3,000 H_2 Balloon Flights During Civil War	1861-64
Dmitrii Mendeleev	Periodic Table of Chemical Elements	1869
Jules Verne	Predicted Hydrogen for a Better Future	1870
John Ericsson	Solar Dish Engines	1872
J. Willard Gibbs	Physical Chemistry; Thermodynamics	1876
Walther H. Nernst	Electrochemistry	1889
Albert Einstein	$E = Mc^2$	1905
J.B.S. Haldane	Solar-Wind-Hydrogen Visionary	1923
Ferdinand Graf Zeppelin	H_2 Zeppelins Safely Flew High and Far	1900-37
Francis T. Bacon	Advanced KOH Fuel Cells	1932-59
Rudolf Erren	Development of H_2 Vehicles	1928-43
U.S. Air Force	Liquid H_2 B-57 Flight Tests	1950-58
John O'M. Bockris	Hydrogen Economy Visionary	1965
NASA	H_2 Moon Trips	1969-72
Roger E. Billings	Hydrogen Homestead Demonstrations	1970s
T. Nejat Veziroglu	Economics of Solar Hydrogen	1990s

The founders and members of the following organizations have made outstanding contributions to The Solar Hydrogen Civilization and I respectfully thank each of them.

International Association for Hydrogen Energy, Dr. T. Nejat Veziroglu

American Lung Association
when you can't breathe nothing else matters.

American Heart Association

Hydrogen Now, Dr. Maurice Albertson

Institute of Ecolonomics, Dennis Weaver

Union of Concerned Scientists

Greenpeace

Home Power

National Hydrogen Association

Association of Energy Engineers

Clean Air Now, Dr. Robert Zweig

American Solar Energy Society

The Sierra Club

National Audubon Society

Rotary International, the Phoenix Rotary Club, and other supporters of the Pollution Free Planet movement as envisioned by Ray Smucker and developed by Craig Wilson, David Wastchak and Bill Chase

Arizona Council of Engineering and Scientific Associations, Charles H. Terrey

Lions Clubs International

Kiwanis International

American Solar Energy Society

World Resources Institute

Federation of American Scientists

World Health Organization

International Solar Energy Society

Earth Policy Institute, Lester R. Brown

Leighty Foundation

International Federation of Red Cross and Red Crescent Societies

Salvation Army Charities

World Church Service

Doctors Without Borders

Most importantly

I thank my long supporting family and founders and members of the American Hydrogen Association for support, inspiration, and determination to provide scientifically proven options for Civilization to overcome the dilemma of dependence upon annually burning over one million years of fossil accumulations. In particular appreciation, I thank:

MY WIFE, KATHLEEN ANN, *our wonderful daughter, Sara Suzann, and her husband, Charles Whitner Enochs, and our extended families especially my parents, T.J. and Lucy Elizabeth McAlister, my brother and four sisters who were taught by T. J. and Lucy that, "We can make it better" in matters ranging from communities to remanufacturing better than "new" engines.*

Present and past Board Members of the American Hydrogen Association

Editors at large of this book

Robert Schafer, for sustainable community development

Bryan Beaulieu, for inventing the hydrogen home living laboratory for sustainable progress

Dr. Lawrence Trimmell, for development of TESI models

Michael Swaine for illustrating this book

Russell Voorhees for urging the publication of this book and expediting this task.

The Philosopher Mechanics and past students of courses that I have taught. Your dedication to conversion of engines for operation on hydrogen will produce the market pull needed to convert the Industrial Revolution to the Renewable Resources Revolution.

TO THE SOLAR HYDROGEN CIVILIZATION

FORWARD

The 1973 energy crisis brought home the fact that sooner or later we would be running out of fossil fuels, viz., coal, petroleum, and natural gas. There were also environmental problems being caused by their utilization, such as air pollution, acid rains, global warming, ozone layer depletion, oil spills, and the surface mining of coal. Some scientists proposed that these twin problems, i.e. depletion of fossil fuels and the environmental problems caused by their utilization could be solved by the Hydrogen Energy System.

In March 1974, at the Hydrogen Economy Miami Energy (THEME) Conference, the Hydrogen Energy System was formally proposed. During the Conference, a small group of scientists, later to be named "Hydrogen Romantics," decided to establish the International Association for Hydrogen Energy (IAHE), to work towards the realization of the Hydrogen Energy System. The Association started organizing the biennial World Hydrogen Energy Conferences (WHECs), and publishing the International Journal of Hydrogen Energy.

Since the formal proposal of the Hydrogen Energy System, there has been much progress. Following the establishment of IAHE, many national Hydrogen Energy organizations were established. Among these was the American Hydrogen Association (AHA) established by Roy McAlister. It was somewhat different than the other national hydrogen energy associations. It was more implementation-oriented and has been very successful at the "grass roots" level, by not only informing the public concerning the benefits of the Hydrogen Energy System, but actually demonstrating these benefits in a laboratory setting, and organizing workshops, courses, and seminars to this end.

Under the leadership of its founding president, AHA has been involved in research and development of hydrogen production, storage and distribution, and utilization of hydrogen. Now Roy McAlister has put together his extensive research and development work in this book, The Solar Hydrogen Civilization. It starts with a description of the present situation, shows why we need Hydrogen Energy, describes how to obtain hydrogen and how to utilize it, discusses the political impediments, talks about the Solar Hydrogen Economy, clears the air on the controversies about hydrogen, and portrays the future with hydrogen, The Solar Hydrogen Civilization. I strongly recommend this book to all those interested in the future of humankind and the planet earth, as well as to the students of the emerging Hydrogen Economy.

T. Nejat Veziroglu
President, International Association for Hydrogen Energy

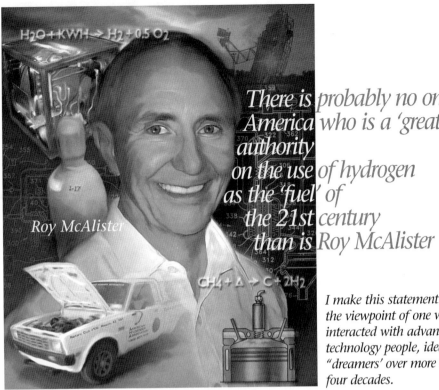

$$H_2O + KWH \rightarrow H_2 + 0.5 O_2$$

There is probably no one in America who is a 'greater' authority on the use of hydrogen as the 'fuel' of the 21st century than is Roy McAlister

Roy McAlister

$$CH_4 + \Delta \rightarrow C + 2H_2$$

I make this statement from the viewpoint of one who has interacted with advanced technology people, ideas and "dreamers' over more than four decades.

Indeed, Roy should be known as *"Dr. Hydrogen."* He has lived and breathed the *"stuff of stars"'* for more than those four decades. He started **the Hydrogen Association in 1964** – nearly forty years ago when he was a young graduate engineer working on a classified government project relating to hydrogen fuel use! *(Perhaps it should be pointed out, while McAlister served as a Research Professor at Arizona State University, and his students routinely won international design-engineering contests with hydrogen projects, his doctoral thesis has never been published as it deals with classified materials which the government has not released to be made public.)* In 1989, the American Hydrogen Association was commenced as a Division of the non-profit Hydrogen Association by a man believing, and fully dedicated, to the idea that renewable energy delivered by hydrogen was/ *is* the best "fuel" for the future.

INTRODUCTION

By Russell Voorhees

Perhaps the best way of providing a concise introduction to Roy McAlister is in the way I did this in a cover letter to over one hundred educators, corporate leaders, scientists and leaders from across the country, a cover letter which was sent along with a draft copy of this book, eliciting response, ideas, and support for the exposure plan for the American Hydrogen Association *(our response from these people was both overwhelming –and gratifying.)* An excerpt of that letter reads as follows:

"………Some ten years ago, when I was involved in a meeting with senior officials of General Electric, Combustion Engineering and General Atomic, *(I was involved at the time with planning the paradigm city for the 21st century –still am)* –for various reasons, I asked "Who is the foremost authority *in the country* on the future power sources and energy production direction –for the next century?"

I was told, "It is Dr. Chauncey Starr" *(President Emeritus of the Electric Power Research Institute in Palo Alto, CA.)*

I met with Dr. Starr. I asked him what the energy source for the 21st century would be. He said, *'It will be hydrogen and methane –for at least the first half of the century.* If the 'interveners' had any idea how serious the problems with nuclear waste are to the environment, they would force a halt to all nuclear production right now. It will take another 50 years for science to control nuclear fusion as a clean source of fuel. Hydrogen and methane are in unlimited supply, and are the energy sources of the future. The challenge is to get past the vested interests and create a transition to these unlimited and clean burning fuels. They are cheaper to produce than fossil fuels. They are clean burning. They are an unlimited source of power/ fuel. They are the future of the 21st century."

As I had already become intrigued with the potential of these fuels, I took Chauncey Starr at his word.

While working with a group I call the *"Disney Graduates"* over the past couple of decades *(the graduates include George Rester –designer of EPCOT center and the proposed experimental city, Kent Bingham –Chief Engineer for the EPCOT project, and now president of Entertainment Engineering ((designers of electro-mechanical projects around the world including the Treasure Island sinking ship in Las Vegas and a multitude of exotic projects, but more particularly "people movers")) –and working with a host of other people of like caliber,)* I became further convinced Dr. Starr was right.

Then, over the past couple of years, I have had occasion to *sit-in* on numerous 'brain-storming' sessions with power and transportation industry leaders. It seems the general conclusion *(and dilemma)* is that Solar/ hydrogen and its sister methane "has no 'clout' at the table of public affairs." If it did, **this nation would already be running on hydrogen**.

With Kent Bingham and Dr. Darrell Pepper *(Dean of the Howard Hughes School of Engineering, UNLV)* I proposed we develop a plan for a credible and highly visible commercial corridor for hydrogen vehicles –to run between Denver and Los Angeles, via Las Vegas. Work has been done. This *'corridor'* has received some national publicity. The problem remains; there is still no spokesman organization with *"clout-and-credibility"* on the National scene, to educate the public.

We always hear, "There are no refueling stations." We always hear, "Isn't hydrogen *dangerous*?" *(result of the Hindenburg fire, a bad rap the fossil fuel industry does not discourage.)* We always hear, **"How can you compete with existing fuel driven industries' influence** in Washington?"

Then I met Roy.
Actually, I had met Roy McAlister some 20 years ago, but a mutual friend insisted I owed it to myself to meet with him again, before starting out on such a program as we were proposing. I had followed the American Hydrogen Association somewhat over the past couple of decades. I had presumed there was neither commitment nor desire to create the kind of public clout necessary on the part of McAlister and the AHA. *I was wrong.*

Roy McAlister is undoubtedly the most knowledgeable hydrogen advocate on the scene today. He started the parent organization of AHA some 35 years ago. He has committed his life to both research and teaching *(Professor, ASU)* in the field, and in development of hydrogen technologies. Roy does not *"toot-his-own-horn"* perhaps as well as others in the field, but he is well-published, including more than 30 patents on hydrogen and related energy conversion topics and is held in very high regard by his peers. Roy has agreed to *"step-up-to-the-bat"* to help Civilization *achieve sustainable prosperity* —and we all need to get behind him, as well as thank him for his contribution to the future of our planet...."

It is past time for the Hydrogen Revolution to take place!

Not believing there was a credible and dedicated organization with the desire and ability to become that credible entity in existence, which could *'take-on'* the education task, we started to investigate ways of developing both public support and Governmental support for the "hydrogen corridor."

The idea of using hydrogen as a fuel source is not new. In fact its use predates the U.S. Civil War *(as you will see when reading this book)* and perhaps should have been the fuel source of preference, had not economics and availability of fossil fuels and potentially huge profits attracted the attention of capital *(money sources)* before hydrogen technologies rose to the point of public awareness. *It is public demand and the flow of money that determines the direction technology may take in our economy.*

The evidence is *overwhelming;* the economics, environmental effects and availability **demand a conversion to hydrogen** as the replacement for *dirty and expensive fossil fuels.*

The *"revolution"* will neither be an easy nor *peaceful* one. Vested interests in fossil fuels and existing related industries have *trillions of dollars at stake*. They will make every effort to protect their investment. They will provide *"dis-information"* while pretending to support alternative fuels. They provide hundreds of millions in contributions to politicians to assure their voice is heard at Federal and State levels. *Roy outlines just a few of his experiences and relates a few of these incidents to support this contention in this book.*

The Solar Hydrogen Civilization is a **must read** for any citizen who wants *some insight as to* how our government operates, who is sincerely concerned with the health of our planet, or wonders how much a gallon of gasoline that we feed through our gas guzzlers of today *really costs*. Roy answers these questions. He also provides data as to the cost to our health and safety the continued *money driven* preemption of alternative fuels actually burdens the voter and taxpayer.

Roy has always been an honest advocate for alternative fuels. In fact, the American Hydrogen Association can be the true source of objective information on fossil fuels, alternative energy sources, and a direction for the *inevitable* new Solar Hydrogen economy.

How did I get interested in supporting Roy and the American Hydrogen Association?

Both as a Patent Attorney at a younger age working with new technologies in the 1960's, and as corporate and community builder in later years, I have watched the evidence mount: *fossil fuels are obsolete* as an energy source. **Environmental considerations demand** a cleaner source of unlimited fuel. Viewed simply, *"The energy of the sun is the source of all needed energy for this planet, and hydrogen is the 'clean' fuel in which this energy can be stored for usage."* Roy explains this much better than can I. He does, in this book.

So, why have a group of *very* knowledgeable **preeminent citizens, educators, engineers and scientists gathered around Roy McAlister and the American Hydrogen Association at this time?** Simply put, we have determined there is no more credible and logical platform from which to launch the "hydrogen revolution" than through the not-for-profit Hydrogen Association, which he founded.

In defense of Executives in the Oil, Utility and related fossil fuel driven companies, it has been reminded many times, "Our legal responsibility is to protect our present investment and maximize profits for our shareholders. While we also have a responsibility to keep abreast of new technologies, we must respond to government or consumer demand. It is the responsibility of the citizens and the government to champion new technologies, and create demand. We will respond."

Roy is known *internationally* for his expertise in hydrogen technologies. Roy has dedicated his life and career to this much-needed *transition* for our economy. For the sake of our *health* as individuals, and for our nation, it is *imperative that* millions of citizens understand the truth about our energy policies. When you have read this book and verified its factual presentations, you will conclude as we did, *we must do something now* to bring about the Solar Hydrogen Civilization. The economics are overwhelming. The positive environmental impact is overwhelming. The positive potential impact on international relations and military cost reductions is overwhelming.

Hundreds of millions of *dollars* are spent each year by fossil fuel people in lobbying our politicians for subsidies and protection of their fossil fuel investments. In reading Roy McAlister's book, one could *almost conclude* fossil fuel industries <u>own</u> a majority of the politicians in America *(he has an interesting quote by Mark Twain along this line in the book.)*

According to data provided in this book <u>*the* **real** cost of gasoline</u> we put into our vehicles is well over $5.00 per gallon right now! We simply are not aware of all the corporate subsidies that are provided from the pockets of <u>middle class taxpayers</u> *(yes, most taxes ultimately fall on those of you who draw a paycheck of less than $100,000 per year, but that is another subject which Roy alludes to in this book.)*

Roy *proves* hydrogen can be produced in volume from Solar and Biomass energy, for as little as $1.00 per gasoline gallon equivalent! As taxpayers we *little understand* the cost of corporate welfare to our national treasury. Nowhere more than the fossil fuel industries, the utilities, and the military industrial complex is corporate welfare more generously dispensed *(unless, to the nuclear industry, which is another story.)*

There was the Industrial Revolution in the early part of the 1800's. There was the Transportation Revolution of the latter part of the 1800's, altered by the automobile, truck, and airplane *'revolutions'* of the 20th century.

There was the "Information Revolution" of the latter part of the 20th century, freeing up the *Internet technology* from the grips of the Military Industrial complex which was the *key* to really launching the period of great apparent prosperity in the 1990's. *(Yes, a group of influential computer executives put their political clout behind Senator Al Gore in the late 1980's. He became Vice President in 1993. The Internet was wrested from the "secret-and-defense" technologies controlled by huge vested interests. Entrepreneurs became able to use this government-developed technology to enhance computer and communications technologies, and profit from this.)* The rest is history, albeit little known history.

With a few million members, *the **American Hydrogen Association** can become the 'voice' for* knowledgeable citizens to educate the public and exert pressure on government(s) around the world to assist in the conversion.

It will take citizens' involvement, as profoundly dedicated as the environmental and Vietnam War *youth movements* of the mid-'60's, to launch the hydrogen economy. <u>*Politicians either listen to money or votes*</u>. This revolution will have to be won with a little money –and a lot of votes. We will have to make the public *aware of the true facts* concerning the Solar Hydrogen economy and its promise for improving civilization. <u>***Roy has done an excellent job***</u> of both laying out the facts, and making the case for the wholesale conversion to Solar/ Hydrogen.

One might ask, what will all the oil and fossil fuels *still potentially available around the world* be used for, if they are no longer needed as an energy source? Roy clearly spells out the many safe and good uses to which fossil biomass should be applied, such as carbon sources for 21st century building materials. Yes, many existing manufacturing plants are, or will become obsolete. The "buggy whip" companies will fight this revolution *tooth-and-nail*. They have already *(you will see this in reading Roy's book.)* They will continue to fight it with all the money, political influence and *dis-information* they can muster. Sure, they will say they are trying alternate energy sources. But, not really to replace fossil fuels. Fossil fuel industries' interests are in staying abreast of new technologies that may become their competitors, and doing their best to control these technologies and protect their present investments. Only when existing fossil fuel and transportation industries see a *trend –and- public demand* for cleaner fuels, which they can no longer fight, will the vested interests *capitulate* and join the transition. Such is the nature of business in a capitalistic economy.

It is up to millions of public-spirited citizens to champion this cause, for the sake of their health, for the sake of the planet, for the sake of our economy, and for the sake of the *very air they breathe.*

Roy McAlister has assigned the rights to this book to the American Hydrogen Association, with proceeds to be used to both educate the public, and to further technology development in the new Solar/ Hydrogen economy.

As we have said in discussions while *"brainstorming"* ways to get the message out to the public and gain support for the hydrogen solution, *"If the American Hydrogen Association could somehow just gain 'clout' comparable to the NRA (National Rifle Association) we would launch the greatest economic revival and era of peace the world has yet seen."* Yes, Congress listens to voter blocks. Sure, it listens most to *payola*, but, **there is nowhere your vote will count more than as an active member of the American Hydrogen Association.**

Comments of others from within our initial support group will be found toward the end of Roy's book *(Appendix.)* All will provide additional *insight* and support for the premises outlined in the book.

Respectfully,

Russell Voorhees,

For the group

As a patent attorney following new technologies, a civic leader involved on a national scale, and as an interested "citizen" Russ Voorhees has been intrigued with alternate energy sources (and following their progress) over the past 40 years. He has been following the career, the research and progress of Roy McAlister over the past 25 years. From this background base, Voorhees has concluded Roy McAlister is the foremost authority on hydrogen and alternate energy technologies in America today. McAlister's accomplishments need to be translated into worldwide applications.

We need Civilization… and it must be made sustainable to conserve the remarkable gains that have been made during the relatively short time of humans on Earth. What is needed now is truthful economic analysis and

THE SITUATION

subsequent accounting to stimulate and justify

improved human performances for much greater economic development than is possible by continuing to be dependent upon burning of fossil accumulations to release energy. Solar Hydrogen is the answer. Humanity has a Grand Purpose. Solar Hydrogen is essential to facilitate Civilization's Grand Plan for Sustainable Prosperity.

The purpose of this book is to provide Civilization with scientifically proven options for solving the difficult and growing problems of economic hardship, international conflict, and environmental degradation that result from dependence upon fossil energy sources. Further, it is to introduce the International Renewable Resources Institute as the suggested way to launch new ventures that are needed to produce the components, goods, and services that enable the Renewable Resources Revolution to accomplish the Solar Hydrogen Civilization and achievement of worldwide prosperity.

So far, Civilization is the greatest human accomplishment. Civilization facilitates individual achievement, transfer of knowledge, and collective accomplishments. It has evolved to serve every real and

CHAPTER 1.0

imaginable need from conception to death and as much more as may be believed. Civilization distinguishes the human condition from every other situation that other life forms have developed.

12

Oil wars have marred the Industrial Revolution. Oil-fueled machines of Hell shrank the world. Petrol powered submarines could sink any surface ship; armored tanks ran over barbed wire and through brick walls; warplanes flew over armies and dropped bombs on strategic targets. Germany wanted to capture land, oil and other resources with WWI. But the War to End All Wars did not solve the problem of expanding demands for diminishing resources. Japan and Germany urgently demanded more oil and resources and caused WWII. Depriving Japan and Germany from access to oil was a successful strategic mission of Allied efforts to end WW II. Struggles to control oil flow from the Organization of Petroleum Exporting Countries (OPEC), other oil rich areas, and pathways from oil fields to market defined the Desert Storm and Afghanistan Wars.

Vice President Cheney was the principal author of the National Energy Policy Report that was released in May of 2001. This report predicts that U.S. requirements for burning 20 million barrels of oil each day will continue to increase. This report predicts that U.S. dependence upon imported supplies of oil will reach two-thirds by 2020 and Persian Gulf countries will be the only sufficient source for this enormous amount of oil. U.S. trade imbalance will continue to grow.

Actually, U.S. dependence upon imported oil could reach 2/3 by 2004 depending upon availability. Otherwise scant petroleum reserves in the U.S. will be depleted more rapidly. U.S. reserves, which were once about as large as Saudi Arabia's, have been mined and burned to the point that we now have less than 3% of the world's remaining oil reserves, but the U.S. burn rate amounts to more than 25% of the world's production.

Iraq has proven oil reserves of 112 billion barrels -- second only to Saudi Arabia. Russia has 49 billion barrels and the Caspian states have 15 billion. Iraq is the only oil producer that can substantially increase oil production to meet growing world demand. Iraq has over $3.8 trillion worth of what the U.S. urges the world to be dependent upon and the price is going up with depletion of world reserves and increasing demand in highly populated countries.

It leaves little wonder why Saddam Hussein generously parceled out production potentials for 44 billion barrels in concessions to Russian, Chinese, and European oil companies. He was trying to use these oil concessions to buy UN Security Council votes against forcing him to disarm for violations of the agreement he entered to end the Gulf War 12 years earlier, and to overcome UN Resolution 1441 requiring him to disarm in 2002. Iraq's concessions of 44 billion barrels amounts to more oil than the total reserves of the U.S., Canada, and Norway which is Europe's highest oil producer.

DOES TEMPORARY SUCCESS OF THE OIL ECONOMY JUSTIFY DEPLETION?

Enormous amounts of coal, oil, natural gas, iron, aluminum, titanium, nickel, cobalt, and refractory metals have been wasted in anger to fuel Earth's oil wars. These resources are wasted in smoke or rotting and corroding hulks on ocean floors beneath battle scenes, along with sacrificed human brilliance and gallantry. Eventually human needs for energy must be met with solar energy and its derivatives such as wind, wave, falling water, and biomass resources. We have opportunities and inevitable responsibility to harness solar, wind, wave, falling water and biomass-waste resources that far exceed any potential for temporary wealth creation by exploitation of oil. Military conflict over oil is beneath human civil dignity and it is shortsighted. We must progress past the Oil Age before we run out of petroleum reserves, and enter a much more important economy of wealth expansion by producing energy-intensive goods and services with renewable energy.

Civilization has experienced exceptional developments during the last 200 years by discovering and rapidly exploiting fossil fuels to achieve enormous productivity. Humans now number some 6.3 billion persons, more than 6-times the largest population that could exist before the discovery of fossil fuels. Even persons with modest incomes in an industrialized country have, in many aspects, much more disposable energy, more comfortable homes, far better appliances, better health care choices, and more enjoyable living conditions than the most wealthy king or queen who reigned before fossil fuels were exploited.

Solar Hydrogen Grand Plan Provides Billions of Jobs to Achieve Sustainable Prosperity

Saving children for a meaningless life of hardship amounts to cruel and protracted child abuse. The 8,000 children who are saved each day by vaccinations against dreaded diseases will need worthy jobs to earn satisfaction that their lives are essential and esteemed in Civilization's accomplishment of the Solar Hydrogen Civilization. It is imperative that far larger, more distributed, and sustainable supplies of solar, wind, wave, falling water, and biomass resources replace present dependency on diminishing supplies of fossil resources. Every person on Earth has a job to accomplish sustainable prosperity and to do less is a form of self-destruction.

1.1

Millions of average citizens regularly drive cars and trucks on improved roadways with sufficient power to comfortably travel 300 miles at 60 MPH. Such powerful freedom to travel and support commerce could not have been matched by the finest horse drawn carriage that existed before the inventions that mechanized the fossil energy revolution. More millionaires and billionaires now exert their economic reigns than at any previous time in history.

Expert accountants and economists commonly consider the cost of a barrel of oil or a ton of coal or a therm of natural gas to be the "taking cost" and the replacement cost is ignored. Disregard of the replacement cost of energy is a serious oversight that causes our sense of value and purpose to be misled. Civilization is threatened with hard times because critical resources are overused and/or misused by the Industrial Revolution.

Illustratively, if the replacement cost of oil had been used to establish the cost of gasoline at the pump, cars would be much more fuel-efficient and air pollution would not plague the cities. If we valued the replacement cost of energy, natural gas would not have been vented to the sky for decades from American oil fields. If the replacement cost of coal had been established in the sale price of coal, central power plants would never have been built that throw away two units of energy for every unit of electricity delivered to customers.

TAKING COST BLINDS MORAL PURPOSE:

1.2 If fossil fuels were priced at their replacement cost, nuclear power plants would never have been pursued because it takes enormous expenditures of fossil fuels to mine, refine, cast, forge, and build all the components of a nuclear power plant. Such enormous expenditures of fossil fuels to commercialize nuclear power could only be ignored because fossil fuels were available at the "taking" cost. If the true cost of nuclear electricity were known in terms of replacement and life-cycle costs it is

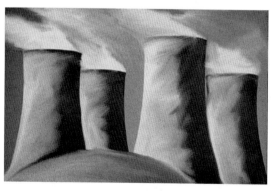

very doubtful that taxpayers would have willingly subsidized the enormous expense to build and operate such energy-expensive plants. Nuclear plants source radioactive wastes that must be guarded against terrorists and disruption of storage provisions by natural earth movements for more than a hundred thousand years. Add these life-cycle costs to the replacement costs for the coal, oil, and natural gas that were expended to build the nuclear power plants and the cost per kilowatt-hour is far more than nearly any other replacement cost of energy.

True economic analysis and accounting must be established that facilitate sustainable development instead of practices that assist in wealth redistribution from the weak or misled by less-scrupulous members of Civilization. Smarter, more productive, harder working individuals will always find ways to prosper but it should not be accomplished by gaining the cooperation of those who are misled by assumptions that the taking cost and not the replacement cost best serves the establishment of our sense of value.

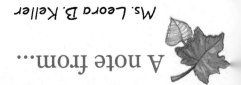

A note from...

Ms. Leora B. Keller

With justice based on truthful establishment of values, Civilization can achieve a grand purpose that unites all races, religions, and creeds in the development of sustainable worldwide prosperity. This Grand Plan for global prosperity will be recognized as a destiny sufficiently worthy to make the sacrifices, exceptional performances, and contributions that are required.

Individuals will find the heart to make the contributions that are needed to fulfill the Grand Plan for sustainable prosperity once this plan is established. History will record the accomplishment of Sustainable Global Prosperity, as Civilization's most worthy and honorable achievement. A renewed spirit for advancing music, art, medicine and other sciences will follow creation of wealth that is measured in environmental improvements along with unparalleled production of Solar Hydrogen energy-intensive goods and services.

SUSTAINABLE MATERIALS and ENERGY:

As Civilization is the greatest collective contribution of humans, Invention is the greatest individual accomplishment. Civilization did not leave the Stone Age because we ran out of stones. Progress past the Stone Age was possible when better tools were invented that were made of bronze alloys. Inventions that emphasized lighter, stronger and potentially more plentiful iron alloys defined the subsequent Iron Age of progress.

New inventions are rapidly progressing that will enable the Solar Hydrogen Civilization to enter the Carbon Age of material excellence. Renewable energy for this advancement will be converted into safely storable, efficiently transportable, and cleanly usable hydrogen. Carbon reinforced products that require less energy to produce and that are ten times stronger than steel, lighter than aluminum, and that conduct more heat than copper will be increasingly used to reduce curb weight of vehicles, improve performance of appliances, and to increase the efficiency of heat-transfer systems. Other forms of carbon will provide super semiconductors and advanced optics. Carbon whiskers, scrolls, tubes, and Buckey balls have launched the Nano Revolution. Hydrogen powered transportation equipment will use compact energy storage tanks made of carbon. Ocean vessels and skyscrapers will be built of exciting new forms of carbon that are much stronger, more corrosion resistant, and that will withstand temperatures that melt steel.

1.3 Suggested timetable for establishing 6,500 new ventures to facilitate and protect sustainable prosperity in virtually every community on Earth

▲ **WITHIN 2 YEARS**

Expand the International Renewable Resources Institute with economic feasibility demonstrations of energy and materials production from renewable resources.

▲ **WITHIN 4 YEARS**

Solicit the best and brightest entrepreneurs, engineers, technicians, accountants, advertising experts, etc., to receive technology transfers and prepare business plans for making, using, and/or selling goods and services that are based upon or that facilitate the advancement of renewable resources. The principal product of the Institute will be business ventures by teams with complementary capabilities that successfully provide products and services to support eventual achievement of the Grand Plan for sustainable prosperity in virtually every community of the world.

▲ **WITHIN 8 YEARS**

Establish appropriate ways and means for preventing catastrophes that threaten Civilization including depletion of one or more essential resources, disease epidemics, collision with a large object from space, depletion of stratospheric ozone and other climate changes.

ENERGY INDEPENDENCE:

Nearly every nation has adequate solar, wind, wave, falling water, and/or biomass waste resources to achieve sustainable energy independence. Even the world's largest energy importer (the U.S.A.) has adequate renewable resources to replace fossil energy sources.

DISTRIBUTED ENERGY PARKS WILL PRODUCE 100 QUADS OF ENERGY:

1.4 Total U.S. energy demand is about 100 Quads per year which is about 100×10^{15} BTU/yr. or 29,308,323,563,892 kWh/yr. Each state can contribute substantial amounts of renewable energy to overcome present dependence upon fossil fuels. Shown below are several approaches that will be developed into a mix of distributed energy production systems. This will provide great improvements in energy security as pollution from fossil fuel use will be eliminated.

SOLAR ENERGY PARKS: Desert areas of California, Nevada, Arizona, New Mexico, Texas, and Colorado receive 6 to 7 kWh/m^2/day in the worst case (January) and 8-10 kWh/m^2/day in June on collection systems with two-axis tracking of the Sun. In a Renewable Resources Park setting that is designed to host tracking dish gensets or tracking concentrators for thermochemical or photovoltaic generators, a spacing of about 4,000 units per mile^2 with 30 to 60 kW capacities can often be achieved. Depending upon the type of power conditioning system selected, this provides 100 to 200 MW/mile^2 of electricity and/or hydrogen generation capacity at daylight times of peak summer demand. The U.S. could become energy independent by building various Solar Energy Parks in California, Nevada, Arizona, Colorado, New Mexico, and Texas that amount to an overall area of about 50,000 to 100,000 square miles or about 1.4 to 2.8 percent of the U.S. territorial area.

Energy for the entire North American continent could be provided from Solar Energy Parks in an area 800 miles long and 150 miles wide along the solar-rich border of Mexico with California, Arizona, New Mexico and Texas. All the energy required by the U.S., Mexico, and Canada could be supplied from such parks and doing so would readily provide jobs for persons that now illegally cross the border for work opportunities. Better jobs will be available in Solar Energy Parks for illegal immigrants who cross the border.

WIND ENERGY PARKS: Wind energy is a form of solar energy. In some areas, wind represents concentrated solar energy because it offers energy densities that considerably exceed the solar radiation reaching Earth's surface. Alaska, Wyoming, California, Nevada, Colorado, Wyoming, Idaho, Arizona, New Mexico, Montana, North Dakota, South Dakota, Nebraska, Kansas, Oklahoma, Texas, West Virginia, and the Great Lakes have locations with excellent average wind velocities. Landowners in Great Plains areas receive royalties of more than $2,000 annually for each 750-kW wind turbine. Modern wind turbines of 1 to 2 MW capacity require a "footprint" of less than 0.40 acres. After allowance is made for "wake cancellation" issues, 30 to 60 such wind turbines can be stationed per square mile. Wind turbines can be placed in Wind Energy Parks at other locations that are not conducive to farming operations to achieve 30 to 120 MW/mile^2 energy production capacity. U.S. energy independence could be achieved by harnessing the wind in 15 to 20 of America's wind-rich areas.

All the energy for Canada, the U.S. and Mexico could be made from distributed Wind Parks along the coastal areas, along with Wind Parks in wind-rich interior areas.

WAVE ENERGY PARKS will harness energy on really wetlands! Virtually all of the vast coastal areas of the U.S. receive energetic ocean waves that represent even more concentrated solar energy than winds provide. Certain coastal areas of Alaska, New England, and Hawaii are particularly fortunate to have very energetic waves. Excellent wave energy potentials are along the Atlantic and

Each megawatt of energy from Renewable Resources Parks will replace fossil-sourced energy that causes the release of about 4,500 metric tons of carbon dioxide along with considerable amounts of ash and heavy metals.

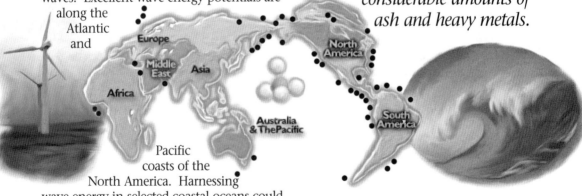

Pacific coasts of the North America. Harnessing wave energy in selected coastal oceans could readily supply 100 Quads of electricity and/or hydrogen needed to achieve sustainable U.S. energy independence.

World offshore Methane-Hydrate deposits.

Wave Energy Parks along Canadian, U.S. and Mexican coasts could also provide all of the energy needed throughout North America.

the Solar HYDROGEN civilization

The Plan

SOLAR HYDROGEN

Our current economy drains natural resources and pollutes the environment. It can be restructured to provide prosperity without pollution by replacing fossil fuels with energy from renewable sources. Hydrogen is the safest, most convenient, and cost-effective way to deliver renewable energy. Hydrogen runs the stars, provides energy for the tiniest to the largest forms of life on Earth, and it can help Civilization become much more friendly and willing to make the appropriate investments needed to achieve future prosperity.

Virtually everyone seeks a better environment. To provide our families with healthful and attractive environments, many of us make large investments to buy homes in landscaped areas with paved streets, modern sewer systems, and city water distribution, along with underground delivery of electricity and natural gas. Vacationers travel around the globe to places with great environments. Life savings are often spent to retire in a better environment.

From this perspective, wealth is a measure of the ability to have a good environment. To a large extent, this ability has been enhanced by the ongoing Industrial Revolution, and this revolution, in turn, has been fueled by various sources of energy. As the Industrial Revolution matured in the early 1930s, anthropologist Leslie White observed that the advancement of civilization is directly proportional to the amount of energy harnessed for each person. White noted that the well-being of each person in modern society depends upon a lengthy inventory of energy-intensive goods and services.

CHAPTER 2.0

Over the past 200 years, our technological revolution has developed exponentially. At the same time, there has been a startling explosion in our global population - from about one billion persons in 1810 to three billion in 1960 to about six billion at the beginning of the twenty-first century. Consequently, the demand for energy has also increased by leaps and bounds.

18

To meet most of our energy needs however, we have been relying on fossil fuels — coal, oil, and natural gas — that are not readily renewable. It has become apparent that the global demand for energy will double in less than three decades and the world's oil reserves could be substantially depleted by then. Long before fossil resources are depleted it will be impossible to match production with demand. Before WWII, Japan, Germany, and Italy needed foreign oil to keep up with growing demands and tried to secure these energy supplies by military actions. In the 1960's the U.S. demand for oil exceeded production capacity and by the time of the oil embargo of 1973 the U.S. was importing 35% of the oil required for the world's leading economy. By year 2003, the U.S. imported 75% of the nuclear fuel, 60% of the oil and 16% of the natural gas used to produce the world's leading economy. Every other country seeks the American standard of energy-intensive life style, but how will it be accomplished?

Future historians will therefore observe that the first few centuries of the Industrial Revolution produced a "technology trap": a trap in which some six billion people annually consumed about a million years' worth of fossil deposits, from reserves that took 500 million years to accumulate. This ominous trap has festered hostility between countries with industrial wealth and those with emerging economies.

As the pace of industrialization around the world quickens, the demand for energy rises in parallel. Considering that our reserves of coal, oil, and natural gas have been seriously depleted in a few decades, it is becoming increasingly urgent to turn to alternative fuels such as hydrogen, derived by utilization of solar and other renewable energy sources.

How can Civilization be safely delivered from this ominous trap?

Our wealth-depletion economy:

2.1 What we have today is a wealth-depletion economy that burns fossil hydrocarbons at a rate equivalent to using 8.6 billion gallons of oil per day. The harder we work in this economy, the less coal, oil, natural gas, and environmental benefits we retain. Yet our definition of "honorable profit" ignores the replacement costs and opportunity costs. (Opportunity cost implies the loss incurred when a resource is used up at lower than optimum value.) This situation is analogous to the bakery that sells cookies without paying the suppliers for flour, sugar, and butter. For a limited time, the proprietor can report great profits, but the business will ultimately close because the suppliers refuse to continue to support the immoral operation.

To understand the opportunity cost in this context, consider the sharp difference in economic benefit between burning fossil reserves as sources of energy and utilizing them for the production of a vast array of durable goods. In particular, 100 gallons of oil can be used to produce polymers that are essential constituents of goods, such as televisions, computers, clothing, carpeting, and vehicle components - that sell in today's marketplace for over $3,500. Moreover, most of these goods can be designed in a manner that allows them to be recycled efficiently into future products. On the other hand, gasoline and diesel fuels extracted from the same amount of oil may yield a onetime income of only $100 or less. This means that for each gallon of oil, gasoline, or diesel fuel we burn, the opportunity cost incurred is more than $34. This already high cost is likely to rise with new inventions that incorporate hydrocarbons in durable goods that improve our quality of life.

In this case, the opportunity cost is hidden from public awareness because oil is generously subsidized to seem relatively cheap and imported petroleum is still plentiful enough to meet Americas current demand for both fuel and non-fuel use. But as demand rises and supplies are depleted, awareness of the opportunity cost will grow and oil will become too valuable to burn. It would, however, be extremely late to start developing technologies to obtain a renewable fuel such as hydrogen at that time. Waiting for widespread hardship and conflict is an immoral opportunity cost.

Burning of Earth's valuable hydrocarbon feedstocks also produces acid-rain constituents, ash particulates, mercury & other heavy metal releases and greenhouse gases such as carbon dioxide. Coal-fired power plants produce 57% of U.S. electricity and emit 90% of the sulfur dioxide and 80% of the oxides of nitrogen from electricity production. Coal-fired power plants produce more mercury emissions than any other source. During the Industrial Revolution, the concentration of carbon dioxide in the global atmosphere has increased by about 25-30 percent. This increase traps more solar energy in the atmosphere, and triggers increased ocean evaporation, more lightning strikes, more rain and snow on the continents, more tornadoes, stronger hurricanes, and other weather extremes.

SEVERE WEATHER TRIGGER:

Water vapor or humidity in the atmosphere absorbs more solar energy and re-radiated infrared energy than any other greenhouse gas. Oceans that cover most of our planet's surface source most of this humidity. Trapping more energy in the Earth's atmosphere and oceans increases the vapor pressure at the surface of the oceans and more water evaporates. Greenhouse gases released by human activities including carbon dioxide, methane, oxides of nitrogen, and halocarbons such as Freon cause increased humidity (water vapor) and increased infrared absorption in the atmosphere. Green plants are invigorated by increased concentrations of carbon dioxide and transpire more water vapor into the atmosphere. Increased concentrations of greenhouse gases are closely correlated to increased severity of weather events. For many persons, additional opportunity costs are measured in injuries and damages to crops, homes, and businesses.

A DEAL THAT WAS MADE
WITH A RADIOACTIVE DEVIL:

2.2

Carbon Sequestration Solution

Methane Dissociation

Carbon

Anaerobic Methane Production

Carbon Durable Goods

H_2 H_2

H_2 Farm Equipment

H_2

BioMass

H_2 Distributed Energy System

Hydrogen Transportation Engines

Green Plants

H_2O

Atmospheric CO_2

The burning of fossil fuels for transportation and other applications rapidly expends natural resources and generates pollutants. It would be wiser to use fossil reserves as starting materials in making the hydrogen infrastructure along with polymers and other substances that are then incorporated into durable goods, carbon reinforced devices for harnessing solar, wind, and falling water resources. This will provide a sustainable future. Sequestering carbon from biomass waste materials to produce durable goods will greatly reduce atmospheric pollution and greenhouse gas accumulations as it stimulates economic development.

2.3

While nuclear fission energy is another source being tapped for the production of electricity, nuclear fuels are also finite in magnitude and could be depleted in a relatively short time. Moreover, dangerous radioactive wastes produced by nuclear fission processes must be carefully contained, inventoried, and guarded against terrorist misuse for hundreds of thousands of years.

Nuclear wastes are accumulating all over the world. In the U.S., dangerous nuclear wastes from electricity plants, aircraft carriers, submarines, bomb factories, missile de-commissioning facilities, and university laboratories are accumulating in 39 states at 131 sites. Two-thirds of the U.S. population lives within 75 miles of one or more of these sites that are all too vulnerable to human mistakes, accidents, and terrorist attacks.

Radioactive waste materials now stockpiled in the U.S. will emit thousands of times more radiation than was released by Chernobyl and many times more than the atomic bombs dropped on Hiroshima & Nagasaki. Radioactive wastes pose difficult technical challenges including the requirement for impervious containment to assure safe transport regardless of all possible accidents and attacks and then provide continuously assured containment for thousands of years along with enormous expense for fail-safe facilities capable of withstanding bombs, earthquakes, floods, and to assure success of armed security personnel for prevention of theft by trespassers and insider crimes.

Plutonium is often one of these radioactive wastes. It will have half the present radioactivity after 380 thousand years of storage. One millionth of an ounce is enough to cause cancer to humans and all other animals that might come into unprotected contact with it. Virtually every agency that has studied the storage issue is aware that the two sites that are supposed to contain nuclear wastes without leaking for hundreds of thousands of years are indeed already leaking. The Idaho National Engineering and Environmental Laboratory and the Hanford Site in eastern Washington are both leaking radioactive substances.

Yucca Mountain, a site about 50 miles north of Las Vegas Nevada has been tunneled and prepared with vaults to receive the nuclear wastes that are now accumulating in 39 states. But experts agree that it is only a matter of time until this site also leaks. Yucca Mountain will eventually have leaks from the designated storage system due to movement of moisture that the mountain accumulates. Like every other geological site, Yucca Mountain is subject to earthquakes and faulting during the many thousands of years required for safe storage.

"We nuclear people have made a Faustian bargain with society ° nobody really knows if we can do this. Trying to project what will happen in thousands of years or tens of thousands of years, is quite ridiculous." This is the analysis of Alvin Weinberg, former director of Oak Ridge National Laboratory where plutonium was tested for atomic bombs that were dropped on Japan to end WWII.

DECEPTIVE REPORTING OF POWER PLANT EFFICIENCIES:

2.4 It has been customary to report the efficiency of generating electrical power as the ratio of electricity produced to the heat energy supplied by burning fossil fuels. On this basis, it is generally held that power plants fired by coal, oil, and natural gas operate at 40 percent efficiency and that utilities are on the verge of achieving 55 percent efficiency in "combined cycle" systems (in which the exhaust from a gas turbine supplies heat for a steam turbine). It is thus suggested that fossil-fueled power plants produce electricity at high efficiency and low cost to consumers.

This calculation is deceptive. It fails to provide fair "apples to apples" comparison with solar-energy options. It is far more meaningful to measure efficiency as the ratio of electricity produced to solar energy required in the primary energy conversion process. Thus electricity is never produced at more than about 0.2 percent efficiency by the best fossil-fueled power plants and almost all electricity is produced at less than 0.1 percent efficiency. (Compared to more than 20% for modern solar dish gensets and photovoltaic panels.)

ONE PART OUT OF 1000 FROM YOUR CENTRAL POWER PLANT:

2.5 Illustratively, a given area of green plants generally converts less than 0.5 percent of the incident solar energy into chemical potential energy, through photosynthesis. Suppose that during the 100 million years from original growth to the present, 50 percent (a generous estimate) of the green plant tissue could be prevented from natural decay and protected by alluvial deposits to become available as coal, which is then burned in a power plant at 40 percent efficiency. The overall efficiency for the conversion of solar energy to coal to electricity would be 0.1 percent [=0.5 (50/100)(40/100) percent]. Similarly, the overall solar-to-electricity conversion efficiency obtained with oil or natural gas appears to be less than 0.05 percent or one part out of 2000.

The harder we work in the fossil-fueled Industrial Revolution, the less clean air and water we will have and as fossil substances are burned we lose the opportunity for using these materials to create a sustainable economy.

It is thus clear that the current ways by which we convert energy are inefficient and costly to the environment upon which all life depends. For a sustainable economy, we need to turn to an alternative fuel that is renewable, non-polluting, and affordable.

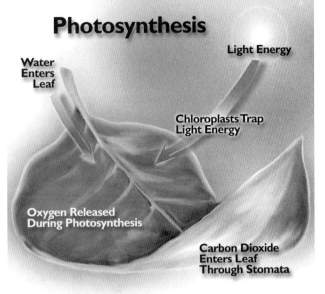

Photosynthesis

Light Energy

Water Enters Leaf

Chloroplasts Trap Light Energy

Oxygen Released During Photosynthesis

Carbon Dioxide Enters Leaf Through Stomata

Water is broken into hydrogen and oxygen. Hydrogen and carbon dioxide are synthesized into plant tissues.

2.6

Green plants such as corn, strawberries, and cotton epitomize wealth expansion by using solar energy to make plant tissues by rearranging atoms from carbon dioxide (from the atmosphere) and water along with dissolved nutrients that are delivered by the roots. Photosynthesis increases Earth's inventory of energy-intensive foods and fibers by adding energy from outer space. In the first step of photosynthesis, solar energy is used to break water into hydrogen and oxygen. Hydrogen is retained to make plant tissues while oxygen is exhaled to the atmosphere, another natural "good" for this hospitable planet.

CIVILIZATION'S GRAND PURPOSE
Creating a wealth-expansion economy:

Hydrogen is a nontoxic, clean-burning fuel that can be produced from readily available hydrogen-containing compounds, such as water, and can be used in practically every application that now use fossil fuels. The term Solar Hydrogen refers to hydrogen that is produced using solar energy or one of its derivative, renewable forms - wind, wave, falling water, and biomass. By replacing fossil fuels with solar hydrogen, we can move from an economy that depletes wealth to one that enhances wealth, health and opportunity for human progress.

The overall efficiency of certain proven technologies that produce hydrogen from solar energy exceeds 20 percent. This is 200 times more efficient than the best overall energy conversion using fossil fuels.

When hydrogen is burned in air, it produces heat and clean water. The heat can be directed to produce electricity or drive an engine; the water may be collected and reused. Two pounds of hydrogen release 122,000 BTU, about as much heat as can be obtained by burning one gallon of gasoline. This amount of hydrogen is therefore referred to as a "gallon of gasoline equivalent," or one GGE.

Methods of producing hydrogen include (a) the splitting (electrolysis) of water using electricity generated from solar energy or a derivative form of energy; (b) thermochemical reactions that release hydrogen from water or biomass; (c) thermal dissociation of hydrogenous compounds using a suitable form of energy such as

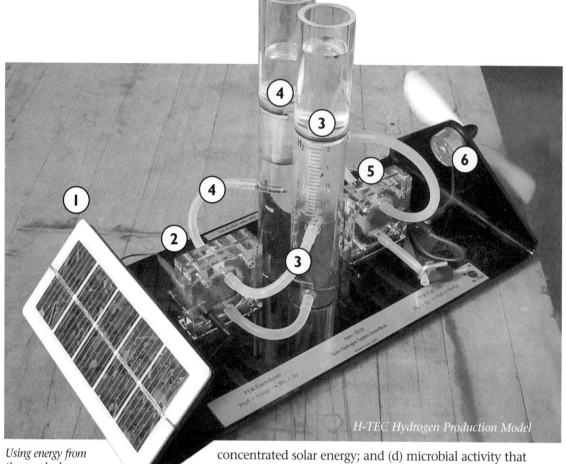

Using energy from the sun, hydrogen can be produced from water by a fuel cell in the regenerative mode. In the fuel cell mode it uses hydrogen to generate electricity. Here, for example, solar energy is converted by photovoltaic panel (1) into electricity that is applied to the fuel cell (2) in the electrolyzer mode to split water into hydrogen (3) and oxygen gases(4). The gases, collected and stored in two cylinders as shown, are fed into the fuel cell (5), which produces electricity that drives a motor (6). In wind-rich areas motor (6) can be operated as a generator to produce hydrogen and oxygen by electrolysis as fuel cell (5) operates in reverse mode.

concentrated solar energy; and (d) microbial activity that releases hydrogen from organic compounds such as biomass wastes.

Solar hydrogen can be used to provide energy for most applications, including transportation, mining, farming, and manufacturing durable goods, as well as in the home. Thus, conversion of the wasteful fossil-fuel economy into the solar-hydrogen economy will facilitate global achievement of sustainable prosperity without pollution.

Transition to solar hydrogen will allow better uses of fossil hydrocarbons. New businesses powered by renewable energy will use fossil feedstocks as raw materials for producing equipment that helps achieve a sustainable economy. For instance, some of these businesses will make modular appliances that use hydrogen to produce electricity and provide hot water and air-conditioning at efficiencies much higher than we have today. Other plants will produce full-size hydrogen-cars that achieve more than 100 miles per GGE, using graphitic materials that are lighter than aluminum and many times stronger than steel, advanced heat engines, and regenerative fuel cells for hybrid applications.

Around the globe, approximately 800 million engines now burn gasoline or diesel fuel and produce health-threatening and property-damaging emissions. These engines and vehicles they power have been built at enormous resource expenditures. A huge parallel investment of energy-intensive resources has been made to build roads, harbors, airports and highways to facilitate travel by these vehicles. One very large market pull for the wealth-expansion economy will be to convert the present fleet of motor vehicles to solar hydrogen.

Hydrogen can be less expensive than gasoline using several production methods. Solar thermal energy can be harnessed to generate less expensive hydrogen by mass production of the appropriate equipment and providing subsidies equivalent to those given to oil, coal, and utility industries. Likewise, given the low cost of producing electricity from existing hydroelectric dams, electrolysis of water could produce lower cost hydrogen. But the lowest cost hydrogen will be extracted from garbage, sewage, farm wastes and forest slash.

Hydrogen holds the world's record for burning faster than any other fuel and at leaner (lower) fuel-to-air concentrations than the fuel

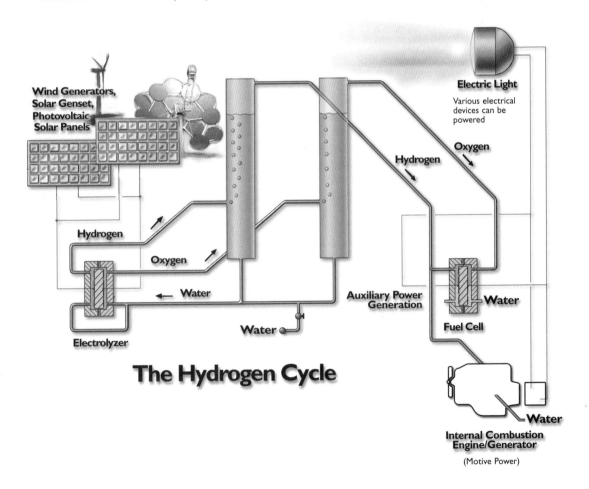

Wind Generators, Solar Genset, Photovoltaic Solar Panels

Hydrogen

Oxygen

← **Water**

Water

Electrolyzer

Auxiliary Power Generation

Hydrogen

Electric Light

Various electrical devices can be powered

Oxygen

Water

Fuel Cell

Water

Internal Combustion Engine/Generator

(Motive Power)

The Hydrogen Cycle

concentrations required for hydrocarbon fuels. These characteristics allow engines that burn hydrogen to operate more efficiently than those that now use other fuels.

The author has proven that virtually every existing engine application from lawn mowers to automobiles, trucks and locomotives can be fueled with hydrogen, and benefits include more power and longer engine life. But perhaps the most astonishing benefit of burning hydrogen in a conventional engine is what we might call *"minus emissions"* - that is, the exhaust gases are cleaner air than the air that enters the engine. Atmospheric levels of carbon monoxide, tire particles, hydrocarbons, pollen, and diesel soot are reduced as air is cleaned by the hydrogen flame. The pollutants are substantially converted into harmless gases, mostly traces of water vapor.

Retrofit kits for high-efficiency combustion of hydrogen can be installed on most engines in the time of a tune-up. Ordinary engines that have been converted to operate on hydrogen show no sign of metal embrittlement or other degradation after decades of pollution-free service. Engine oil stays clean, spark plugs last much longer, and degradation as measured by corrosion and wear on piston rings and bearings is greatly reduced. These engines can therefore start easier, last longer, run better, and clean the air. And this transition will facilitate the distribution of hydrogen for the advent of fuel cells on a commercial scale. Several major auto manufacturers have successful experimental fleets of hydrogen-powered vehicles.

Solar hydrogen can make any country energy-independent and pollution-free as far into the future as the sun will shine. Illustratively, just a small portion of North America can produce enough solar hydrogen to supply all the energy needs of the whole continent. In this manner, if solar hydrogen is used to provide energy-intensive goods and services to the world's population, it will facilitate wealth addition as opposed to wealth depletion that results from burning fossil resources. The harder we work in the solar-hydrogen economy, the more goods and security everyone can have, resulting in a lower rate of inflation and less impetus for conflicts and strife. These circumstances create global incentives to work for higher sustainable living standards, for both present and future generations.

Throughout history, civilizations have flourished when a "grand purpose" was established, and faltered in the absence of serving a purpose that could continue beyond the lifetimes of individual participants. The Egyptians had pyramids, the Greeks had democracy, Romans had conquest, and the Industrial Revolution is based on our quest for increased productivity. But the value of the Industrial Revolution is diminished as long as it is beset by problems such as environmental pollution and depletion of limited resources.

Now that we know of cleaner, healthier, sustainable, and more profitable ways to make a living, it is foolhardy to burn fossil deposits and pollute our planet. We have a moral responsibility to ourselves and future generations to replace fossil fuels with appropriate renewable energy selections.

A worthy grand purpose, therefore, is evolution of the Industrial Revolution into the Renewable Resources Revolution. It will improve the environment and allow civilization to escape from the trap of dependence upon fossil fuels.

Recently, the Voyager spacecraft left our solar system on a one-way journey into deep space. As it reached the outer limits of our solar system, this traveling telescope turned toward Earth and took a picture. Even from the relatively short distance of 4.5 billion miles, Earth is hard to find in this high-magnification image. In the vastness of space, Earth is a minuscule speck, a seemingly unimportant blue dot that takes a year to orbit around a relatively minor middle-aged star.

Civilization must become much more aware of the perspective of this faint blue dot in the vast vacuum of space. It is by countless miracles during the billions of years before human appearance on Earth that we now have a benevolent environment to support us. We must do everything possible to protect this environment, including its diversity of life-forms that we depend upon. We must dedicate ourselves to defining and achieving the *grand purpose of civilization* and attain *sustainable prosperity without pollution*.

KEY TECHNOLOGIES

An economy that is powered by solar hydrogen will require a technological base in the areas of production, storage, distribution, and utilization of hydrogen and electricity from renewable energy sources. This base could take the following form:

1. Establishment of Renewable Energy Parks, where various types of generators convert solar energy and its derivatives (such as wind, falling water, wave motion, and biomass resources) into electricity and/or hydrogen. For instance, some of the parks will include solar dish "Gensets" (with parabolic solar concentrators) that hold the world's efficiency record for converting solar energy to electricity. Other parks will have wind generators for the same purpose. Still others will convert sewage, garbage and farm wastes into methane, carbon and hydrogen.

2. Storage of surplus hydrogen and methane in depleted natural gas and oil wells and similar geological formations.

3. Utilization of the existing infrastructure of electricity grids, natural gas pipelines, highways, and rails to distribute methane, hydrogen and electricity from renewable resources. For instance, locally produced hydrogen can be added to natural gas in existing pipelines and be selectively removed by filtration at desired locations.

4. Installation of kits that enable the world's current population of internal combustion engines and motor vehicles to operate with directly injected hydrogen, and/or renewable methane. A vehicle

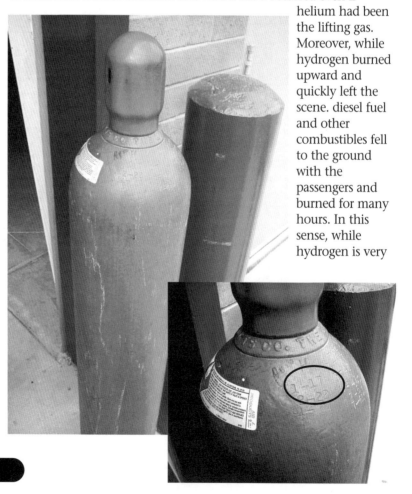

converted to operate on hydrogen can clean the air through which it travels and still revert to gasoline if needed.

5. Installation of hydrogen fuel cells in place of alternators and lead-acid batteries to allow hybrid engines to be much more efficient and to provide better performance.

6. Introduction of vehicles that run on hydrogen-based fuel cells in place of internal combustion engines.

7. Installation of reversible electrolyzers (fuel cells) in homes and businesses. With these devices, surplus electricity during off-peak hours can be converted to storable hydrogen, and conversely, stored hydrogen can be used to make electricity to meet peak demands.

Handling Hydrogen Safely

The tragic accident in which the Hindenburg was consumed in a fire in 1937 still haunts the public memory. It conveys the impression that hydrogen, the gas that filled the dirigible's lifting cells, is too dangerous for public use. Actually, NASA investigators have concluded that the fire was started by the ignition of flammable surface materials and would have occurred even if helium had been the lifting gas. Moreover, while hydrogen burned upward and quickly left the scene. diesel fuel and other combustibles fell to the ground with the passengers and burned for many hours. In this sense, while hydrogen is very

Steel tanks that were first hydrostatically tested in 1917 for storage and transport of hydrogen. Twenty years before the Hindenburg fire these tanks were on the American roads and have continued to pass every test for safe delivery of hydrogen.

CARBON REINFORCED TANK SURVIVES SEVERE CRASH:

This series of photographs show a crash impact test of a vehicle with a hydrogen and/or methane storage tank. (1) Stripes were painted on the rear portion of the vehicle to show what happened when the vehicle hit the ground to simulate a rear end collision. (2) A crane hoisted the car to an altitude that produced a free-fall velocity of 60 MPH at ground level. (3) The hydrogen and/or methane tank is located in the trunk of the vehicle near the rear window. (4) During the crash the rear portion of the car progressively crumpled. (5) Major portions of the vehicle were crushed, the gasoline tank was ruptured and the engine was torn from its motor mounts. (6) Appearance of the hydrogen and/or methane storage tank that continued to safely contain hydrogen and/or methane at 3,000 PSI after the crash impact test.

Appearance of the Hydrogen or CNG tank after a 60 mph crash test as the rear of the car crushed around it, leaving it intact and safe for continued storage of hydrogen.

flammable, it is less dangerous than gasoline and other fuels that are heavier than air.

Since 1917, millions of similar steel tanks have proven that compressed hydrogen can be stored and shipped countless hundreds of million miles at more than 2,000 PSI (pounds per square inch pressure) and be safer in virtually every circumstance than gasoline.

These 85 year-old storage practices show no sign of degradation of the steel from long-term exposure to pure hydrogen. Hydrogen can also be stored in many other ways, including as a super-cold solid at 13.8K (-434.8°F), a cryogenic liquid at 20.4K (-423.0°F), a hydride compound with active metals, a dissolved gas in other materials, or as an adsorbed gas on activated carbon. Transoceanic delivery of liquid hydrogen by cryogenic tankers is comparable to delivering liquid natural gas, which is an established business.

Currently, hydrogen can be stored safely in tanks reinforced with carbon fibers that are 10 times stronger than steel. These fibers can be produced from waste hydrocarbons in processes that are powered by solar energy or its derivatives. The composite tanks readily resist the impact of a 100-MPH collision, an attack with a .357 magnum pistol, or a bonfire test in which the tank's surface reaches 1,500°F.

Pipeline Transportation Infrastructure

Steel tanks that have held hydrogen for more than 85 years show that modern steel pipelines now carrying natural gas can readily transport mixtures of renewable hydrogen and methane along with natural gas. This facilitates an important opportunity to link producers of hydrogen from solar-rich, wind-rich, falling water and wave-rich areas with markets in far away cities that are presently served by natural gas.

ABSOLUTE
Storage

Contrary to the common belief that steel will not hold hydrogen, this old steel tank has held hydrogen for more than 85 years, showing that steel pipelines that now carry natural gas can readily transport mixtures of renewable hydrogen and methane along with natural gas. This means that producers of hydrogen from solar rich, wind rich, and wave rich areas can bring their product to markets in far away cities that are presently served by natural gas.

THE SUSTAINABLE FUEL
HYDROGEN
AMERICAN HYDROGEN ASSOCIATION

METHANE... FRIEND OR FOE?

Converting Greenhouse Gases into Profitable Products

Polar snow core records show the relative concentrations of atmospheric gases for the last 160,000 years. Present levels of methane are more than two times higher than at any time in the polar snow core records. Halogenated hydrocarbons only appear in snow layers of the last 60 years. Carbon dioxide levels are up about 30% compared to highest previous levels in polar snow records.

Each methane molecule is about 27 times more harmful as a greenhouse gas than each carbon dioxide molecule. Methane is released into the atmosphere by anaerobic decay of organic matter including digestive processes of organisms ranging from bacteria to termites and mammals, by volcanic pyrolysis of heavier hydrocarbons, and by venting of oil and natural gas formations. Tables 3.1 and 3.2 show atmospheric greenhouse gas concentrations and relative impacts.

CHAPTER 3.0

TABLE 3.1 **Heat Trapping capacity**

References: Henning Rodhe, "A Comparison of the Contribution of various Gases the Greenhouse effect", Science Vol 24-8, June 1990. Ontario Energy Educators (1993)

ATMOSPHERIC SPECIES	RELATIVE HEAT TRAPPING EFFECT	DECAY TIME (YEARS)
CO_2	1	120
CH_4	27	10
N_2O	200	150
CFC-12	10,000	120

TABLE 3.2 **Comparisons of Greenhouse Gas Impact**

References: Henning Rodhe, "A Comparison of the Contribution of various Gases the Greenhouse effect", Science Vol 24-8, June 1990. Ontario Energy Educators (1993)

Species Concentration (PPBV)*		Rate of Increase (% PER YEAR)	Contribution (RELATIVE % OF TOTAL)
CO_2	$= 353 \times 10^3$	0.5	60
CH_4	$= 1.7 \times 10^3$	1.0	15
NO_x	$= 310$	0.2	5
CFC -12	$= 0.48$	4.0	8

*ppvb = parts per billion by volume

Most oil field development and production starts with considerable methane release from oil that is de-pressurized as it reaches the surface. Because it is more expensive to build pipelines to transport methane to market than oil, methane is typically wasted by venting it into the atmosphere or burning it. In the early days of most American oil fields it was common to burn natural gas and send the oil to market by tanker trucks and then by pipelines as they were installed.

Methane was burned in such quantities in oil fields throughout Texas, Oklahoma, and Kansas that a newspapers could be read at night by the light of methane torches that were wasting this valuable resource. Eventually federal subsidies were provided to collect much of the natural gas being wasted and America now has an extensive natural gas delivery grid that serves the great majority of the U.S. population. In Russia and other countries with enormous natural gas reserves that are relatively remote, large amounts of methane have been wasted in more recent times.

DOMESTIC SOURCES OF METHANE:

Garbage and sewage have presented disposal problems throughout the history of civilization. Offensive to our senses of sight and smell, garbage and sewage wastes have been discarded in places that are out of sight and far enough from where we live to spare us from the repugnant odors of rot and decay. Eventually wet sewers were developed for more convenient transport of wastewater and sewage to local rivers, lakes, or to the ocean.

3.1

Beginning in 1619, supposedly "good" water was piped into the homes of London. This "city water" provided much more convenient and eventually safer water than carrying a bucket at a time from the river or a local pump. Regular bathing, clothes washing, and floor scrubbing became more practical and expected of self-respecting citizens. Hoping for better sanitation and to eliminate the fear of something bad falling on you, King James I issued a building code prohibiting the use of overhanging commodes and the dumping of chamber pots into the city streets. Thomas Crapper was eventually knighted for his invention of the modern flush toilet and the new use for city water that it created. Mr. Crapper called his invention the "water closet" and it used city water to flush unmentionable wastes out of sight and out of mind.

Wet sewers and flush toilets increasingly became the standard for city dwellers and it became apparent that rivers could not sustain the growing load of decaying anthropological biomass without becoming unbearable and dangerous sources of disease. Sewage disposal plants were developed to kill dangerous microbes and remove concentrated sludge that could be buried in landfills, barged out to sea, or used in other applications. The water remaining after the sludge removal processes could be drained to rivers, lakes, and oceans with less danger.

As the industrial revolution sourced more and more ready-to-eat, ready-to-use, appeal-packaged disposable products, the scope of garbage accumulation and disposal became enormous. Giant landfills have become the standard place to dispose of daily arrivals of garbage and sewage sludge carried by truck caravans.

Wastes are hauled to remote areas, compacted by tractors into trenches and covered with soil. But this approach generally produces continuing problems including contamination of ground water, the stench of rotting biomass, and a trail of paper shreds and trimmings from trees that litter the roads to the dump and the area around the dump.

Eventually standards developed to assure public safety by requiring garbage collection trucks to have provisions for compacting or at least containing garbage being transported to landfills. Landfills are required to seal wastes such as garbage and sewage sludge within an impermeable liner and to provide for positive venting of gases such as carbon dioxide and burn methane with provisions for preventing the escape of poisonous hydrogen sulfide fumes.

Bothersome Emissions:

During the last 100 years, it has been increasingly observed that greenhouse gases are accumulating in the atmosphere. Even the most modern sewage-treatment and garbage disposal facilities eventually release the carbon and hydrogen to the atmosphere as methane (CH_4), water vapor (H_2O) and carbon dioxide (CO_2). If the disposal conditions are provided with an abundance of air, the releases of carbon and hydrogen are predominantly as carbon dioxide and water vapor

If the disposal conditions are anaerobic (in the absence of air or oxygen) the carbon and hydrogen releases shift to methane (CH_4) or possibly hydrogen (H_2) depending upon the kind of garbage, type of microorganisms present, and conditions of disposal including the temperature, pressure, acidity and amount of water mixed with the biomass. Landfills designed to be anaerobic digesters generally produce 40% - 60% methane, about 40% carbon dioxide, and much less hydrogen sulfide, nitrogen, water vapor, and various other gases.

These gases escape into the atmosphere and allow much of the solar radiation including ultraviolet, visible, and infrared wavelengths to penetrate to the surface of the earth. Upon reaching Earth's surface waters and soils, these solar wavelengths that represent the 10,000°F radiation spectrum of the surface of the sun are converted to a much longer-wave spectrum (-60° to 120°F) as radiation is converted into heat. Greenhouse gases such as carbon dioxide, methane, and water vapor in the atmosphere absorb the longer wavelength infrared spectrum being radiated from the earth's surface. This provides the *greenhouse gas radiation trap* for heat energy that would have escaped into the cold black vacuum of space.

Many of the greenhouse gas constituents are man-made molecular species that have been produced in huge quantities. If water vapor is omitted from comparison, carbon dioxide collects about 60% of the energy in the atmosphere. Methane (CH_4) is responsible for about 15% of the greenhouse gas energy collection in the atmosphere. Chlorocarbons and chloroflurocarbons (CH_3Cl, CFC, Etc.) collect about 8% and oxides of nitrogen NOx (NO, N_2O, Etc.) collect about 5%. (See Tables 3.1 and 3.2)

Since greenhouse gases cause more heat energy to be trapped near the Earth's surface this results in increased evaporation of the oceans; more extremes in the weather, including tornadoes, hurricanes, snow or ice storms, and flood-producing rains; and the possibility of a trigger effect to cause catastrophic warming of the atmosphere. The experiment that is being conducted to see what happens when enormous amounts of greenhouse gases enter the atmosphere is without parallel in human history. Never before have there been as many lives at stake, economic progress at risk, and never before has such a rapid increase in new types of greenhouse gases been provided.

IMPRISONED TRACE MINERALS and NUTRIENTS:

Nutrients such as calcium, phosphorous, potassium, iron, magnesium, zinc, iodine, vanadium, cobalt, molybdenum, manganese, chromium and nitrogen, along with trace minerals are continually extracted from the land and forest soils. An average of 1,646 pounds of garbage per U.S. citizen is produced each year. This amounts to about 231,900,000 tons of waste containing various amounts of essential minerals and nutrients that are taken from productive soils and forests. Most of the mass of garbage and sewage consists of organic forms of carbon and hydrogen that are supplied by ground water and air but the soil supplies the mineral constituents without which the plant could not grow and produce plant tissues including the crops taken to feed and clothe city dwellers.

3.2 Adding greenhouse gases to the atmosphere has an effect similar to adding more fuel to an engine. With additional energy the engine does more work. The global atmospheric engine is doing more work and we record the results as increased occurrences of damage due to severe weather events.

Sewage sludge and garbage continue to decay after being buried in landfills and produce greenhouse gases such as methane, carbon dioxide, and hydrogen sulfide. In addition to causing unwanted releases of carbon dioxide and methane, landfills and organic wastes dumped into the oceans effectively prevent essential soil nutrients such as fixed nitrogen and trace minerals contained in such wastes from being replenished as needed for productive farmlands and forest soils.

After these essential soil nutrients are taken to landfills as sewage sludge and garbage, they may remain imprisoned for hundreds or thousands of years with little or no intention of ever being restored to the farm soils that sourced them. This is an unfortunate trend and is largely overlooked because crops seem to grow well when provided with commercial fertilizer and irrigation water. Even when large crop yields are produced, however, they may have questionable nutritive values. Foods from depleted soils have much lower levels of trace mineral values than the crops produced when native prairies were first plowed and planted.

Fugitive Emissions: captured for Profit:

New ventures are being developed for collecting, processing, and distributing renewable energy from biomass landfills. Shallow wells are placed in landfills for low-pressure extraction of methane. These wells produce renewable methane and other gases such as carbon dioxide, nitrogen, water vapor and hydrogen sulfide. This mixture of raw gases is collected from the landfill discharge vapors.

In order to increase the energy value per volumetric measure, methane is filtered through special membranes and may be liquefied to facilitate delivery by cryogenic trucks or rail tankers to market. This filtration and/or liquefaction process eliminates particulates and gases such as carbon dioxide and nitrogen, which do not provide heat upon combustion. Finished products may be separate streams of very high quality liquefied carbon dioxide and liquid methane with more dependably constant energy content, viscosity, and dew point compared to pipeline supplies of conventional natural gas. In vehicular applications, liquid methane also avoids contamination and plugging of fuel injectors with compressor oil that often plagues compressed natural gas systems.

In comparison with natural gas, which may require transport through thousands of miles of pipelines and many compressor stations to connect distant gas fields with markets, delivery of landfill methane to local markets requires much less pipe and only one compressor station which is used in the liquefaction and/or separation process.

Natural gas varies in composition from well to well and with time over the life of a producing oil or natural gas well. Table 3.3 compares combustion characteristics of natural gas with methane and several other fuels. In the early life of a new oil well, the natural gas mixture may contain considerably more butane, propane and other heavier hydrocarbons and as the reservoir is depleted the methane content increases.

As shown in Table 3.4, methane is the major ingredient of natural gas, which typically contains lesser percentages of heavier hydrocarbons such as ethane, propane, and butane. Because these

TABLE 3.3 Fuel Combustion Characteristics

Fuel	Lower Flame Limit	Upper Flame Limit	Lower Heat Release (BTU/lb)	Higher Heat Release (BTU/lb)	Air-Fuel Ratio	Flame Speed* (FT/Sec)
Hydrogen	4.0% vol	75.0% vol	51,593	61,031	34.5 lbs/lb	30,200
Carbon Monoxide	12.	74.2	4,347	4,347	2.85	
Methane	5.3	15	21,518	23,890	17.21	4,025
Ethane	3	12.5	20,432	22,100	16.14	4,040
Propane	2.1	9.4	19,994	21,670	15.65	4,050
Butane	1.8	8.4	19,679	21,316	15.44	4,060
Benzene	1.4	7.1	17,466	18,188	13.26	4,150
Methanol	6.7	36.5	7,658	9,758	6.46	3,900
Ethanol	3.2	19	9,620	12,770	8.99	4,030
Octane	1.1	7.5	19,029	20,529	15.11	4,280
Hexane	1.18	7.4	19,017	20,419	15.1	4,200
Gasoline	1.0	7.6	18,900	20,380	14.9	4,010

*At Atmospheric Pressure ** Higher Heat Release Value includes the heat of condensation, as steam becomes liquid water

heavier hydrocarbons occupy less space than methane there may be slightly more energy in natural gas than in methane.

Because "natural gas" is any combustible mixture that comes out of the well from petrocarbon deposits, it may contain varying amounts of methane, ethane, propane, butane, casing-head gasoline, hydrogen, and non-fuel constituents such as water vapor, nitrogen, carbon dioxide, hydrogen sulfide, and helium. Coal-seam deposits produce natural gas that is higher than average in methane. Some wells around Junction City, Kansas, produce high percentages of hydrogen. Many Pennsylvania wells are known to produce "rich gas" that is high in ethane, propane, butane, and natural gasoline.

In order to standardize the quality of natural gas that reaches customers though pipeline delivery, the heating value must exceed 900 BTU on a lower heating value. This means that upon complete combustion, 900 BTU/scf must be released before the water that is produced by combustion is condensed and cooled to standard pressure and temperature. Poisonous and corrosive constituents such as carbon monoxide and hydrogen sulfide must be removed and an odorant such as mercaptan must be added. Water vapor must be reduced to a concentration that prevents ice from building up in valves and exposed lines in cold weather.

Filtering out condensable ethane, propane, butane and water can reduce these cold weather complications. Another approach is to compress, cool, and liquefy methane and use the liquid methane as a compact cryogenic fuel or as compressed gas following heat addition and re-gasification.

Comparison of Typical Natural Gas and Methane   TABLE 3.4

Pipeline Natural Gas	Pure Methane
I scf = 923.7 BTU (LHV)	I scf = 912 BTU (LHV)
I scf = 1,000 BTU (HHV)	I scf =1008 BTU (HHV)
I lb = 21.826 BTU (HHV)	I lb. = 23.635 BTU (LHV)
I GGE* = 5.66 lbs.	I GGE = 5.309 lbs

GGE means gallon of gasoline equivalent

In addition to cost savings, collecting biomass methane will greatly reduce atmospheric pollutants. Gasoline combustion would produce about 20 pounds of fossil-age carbon dioxide per gallon of gasoline. By substituting methane from landfills for gasoline, there is no new carbon dioxide and the benefit can be considered to be another minus-emissions bonus because fossil carbon from the combustion of gasoline is replaced with biomass carbon. Carbon dioxide that comes to green plants though the air is photosynthesized into plant tissue. When the plant residues are converted to methane in landfills and recovered for use as an engine fuel, there is no net addition of carbon dioxide.

Collection of landfill methane prevents its escape as a greenhouse gas. This is a less than zero emissions result for the combination of fuel source and vehicular use. A much greater minus-emissions result is accomplished when the methane is dissociated into carbon and hydrogen and gasoline is replaced by hydrogen.

After extraction of fuel gases, soil nutrients can be efficiently collected, transported, and utilized as fertilizers. This is encouraged by the fact that extracting energy constituents such as carbon and hydrogen from biomass virtually eliminates dangerous microbes and most toxins found in biomass wastes.

For the vehicle operator, the price of renewable hydrogen, methane or Hy-Boost fuel at the compressed gas pump can be lower than gasoline. For the city manager struggling to comply with clean-air rules, the combined pollution emissions from municipal vehicles and waste disposal facilities are greatly reduced in comparison with present practices. As shown in Tables 3 and 4, it is twenty- to seventy-times better to use methane as a fuel to produce carbon dioxide than allowing the methane to escape from wastes. It is thousands of times better to convert methane to hydrogen and carbon enhanced durable goods and to replace gasoline with such hydrogen. Additional greenhouse gas reductions result from displacing fossil fuels.

Methane is not poisonous to persons exposed to the refueling operation compared to the highly poisonous and carcinogenic ingredients of gasoline, especially gasoline with oxygenation additives such as Methyl Tertiary Butyl Ether (MTBE) and methanol. For the economist, values are created from wastes that would have rotted into the atmosphere.

DANGEROUS GAS MONSTERS LURK BENEATH A STRANGE ICE:

Modern tillage practices and the use of fossil-based energy intensive fertilizers have provided enormous increases in agricultural productivity. In many instances however about four pounds of topsoil erode into streams and rivers for every pound of food that is produced by crops. Topsoil contains significant amounts of organic substance. Earth's rapidly growing human

3.3

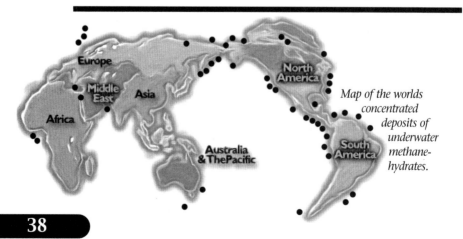

Map of the worlds concentrated deposits of underwater methane-hydrates.

*I*t is twenty-seven times better to use methane as a fuel to produce carbon dioxide than allowing the methane to escape from landfills. It is thousands of times better to convert methane into carbon and hydrogen and to replace gasoline with this renewable hydrogen.

TABLE 3.5 **Fuel Characteristics**

PROPERTIES UNITS	HYDROGEN	METHANE (NATURAL GAS)	GASOLINE
1. self-ignition temp, (C)	585	540	228 - 501
2. Ignition limits in air (Vol%)	4 - 75	5.2 - 15	1.0 - 7.6
3. Min. ignition energy (mWs)	0.02	0.29	0.24
4. Flame speed (cm/s)	265	40	40
5. Flame temperature (^{0}C)	2045	1875	2200
6. Lower heating value (kWs/g)	120	50	44.5
7. Detonation limits (Vol.%)	13 - 65	6.3 - 13.5	1.1 - 3.3
8. Detonation velocity (km/s)	1.48 - 2.15	1.39 - 1.64	9.4 - 1.7
9. Expl. Energy (kg TNT/m^3)	2.02	7.03	44.22
Diffusion coefficient (cm^{2}s)	0.61	0.16	0.05

population demands increased food and fiber production. In many areas of the world there is a shortage of cooking and heating fuel. Surface vegetation has been stripped and burned for cooking and heating. This exposes soil to erosion. Exponentially increasing demand for food, fiber, and fuel has hastened the rate of soil erosion, and organic soil materials are increasingly washed into streams, rivers and oceans.

Much earlier in Earth's geological history, the Gulf Mexico reached north to the areas now known as Kansas and Nebraska. Continental lifting and erosion of mountains and higher topologies of the continent provided the fill to eventually push the ocean south to present Gulf Coast areas of Alabama, Louisiana and

Texas. Continuing continental erosion carried silt through the creeks and rivers to the swamp deltas and beyond to the deeper portions of the continental shelf. Substantial deposits of organic materials were covered with sufficient alluvial depositions to eventually produce much of the coal, oil, and gas now found in formations that are now buried beneath thousands of feet of overburden.

Similar soil erosion results in depositions of organic materials as sediment in anaerobic ocean waters around every continent. In areas that are sufficiently deep and cold, methane produced by microbial induced anaerobic decay of organic matter is retained as crystals of methane molecules within surrounding cages of water molecules. Called "clathrates" or "methane hydrate" when the enclosed gaseous portion is predominately methane, this ice is marginally stable at ocean depths of about 260 meters (850 feet) and deeper, depending upon the type of silt and local temperature.

Clathrates can store surprisingly large quantities of trapped gases. A cubic foot of methane hydrate with 100% packing efficiency of the methane in the clathrate cavities could hold about 164 cubic feet of methane gas if released and measured at standard pressure and temperature. Most of the world's ocean deposits, however, occur as mixtures of silt and methane hydrate and contain about 100 cubic feet of methane per cubic foot of methane hydrate. (1)

At 32°F and 382 psia (corresponding to 262 meters or 861 feet deep) the ratio is 6.85 molecules of water per methane molecule (2). At higher pressures (greater depths) the methane hydrate is stable at higher temperatures. If the pressure is released or the temperature increases beyond the stable temperature, the ice decomposes and releases the caged methane. As the height of gas methane crystals and sediment increases it adds insulation over the lower portions to cause the temperature to rise as heat diffusing upward from the Earth's core is retained.

1. William P. Dillon, U.S. Geological Survey, Encyclopedia of Physical Science and Technology, Third Edition, Volume 6, (2001)

2. Kirk-Othmer Encyclopedia of Chemical Technology, 2nd Ed., Volume 13, pp.364.

DOES IT FLOAT OR SINK?

3.4 Ice made of water is less dense than water and it floats. Methane hydrate is also slightly lighter than water at ambient pressure but does not float until it is released from the bottom where it formed within sediments that are heavier than water.

After a methane deposit reaches certain thickness and consequent insulating capacity, the lower layers retain heat and warm sufficiently to decompose. Methane is released and builds a pressurized gas layer in the lower zone due to the low permeability that is imposed by the over burden of silt and methane hydrate deposits along with the pressure of salt water above. This causes a slower speed of sound propagation at the interface of collected gas methane hydrate and it produces a distinct reflection of seismic signals that characterizes the lower boundary of the methane hydrate deposit. Seismic mapping of methane hydrates has been rapidly accomplished and comprehensive indications of the locations and amounts of methane hydrates in the Earth's oceans are known.

Methane hydrate deposits in the oceans are not protected by

the thousands of feet of stable salt deposits, shale, limestone, and other strata usually found over natural gas and oil deposits on the continents. Methane hydrates often contain extensive methane gas layers that are merely trapped below a weak layer of mixed sediments and methane hydrates. Increases in temperature and/or reduction in the water depth and resulting reduction in the pressure causes these hydrates to decompose and release more trapped methane. Seismic disturbances and volcanic activity undoubtedly cause these marginally stable methane hydrates to release methane from gas pockets and/or by decomposition of the clathrate ice.

During glacial periods the sea level drops as water is retained on the continents as snow and ice. This may indicate a cycle in which carbon is tied up in methane hydrates and other carbonaceous deposits until the atmosphere cools sufficiently to create an Ice Age that ends when the sea level declines sufficiently to release the methane and cause greenhouse gas warming. And it begs the question: How much climate change would it take to seriously jeopardize the farming operations that now barely support Earth's growing human population?

MINING EFFORTS MAY RELEASE DANGEROUS BUBBLES THAT EAT SHIPS

Figure 3.5 shows the hazard to a ship that would be suddenly and mysteriously sunk by encountering a rising bubble of methane that offers even less buoyant support than air. It has been suggested that some of the ships that disappeared mysteriously in the Bermuda triangle may have been sunk as methane bubbles suddenly emerged from below and capsized them. Similarly, tales of mysterious sea monsters that swallowed ships might be founded on observance of a ship that suddenly disappeared in a calm sea... in a rising methane bubble.

The probability of a ship passing the location that a methane bubble might escape from a destabilized clathrate deposit is relatively small. A much more widespread and significant danger stems from dangerous global weather

Civilization must face the danger and realize that converting methane hydrates into carbon products and hydrogen is actually an urgent opportunity that must be taken with far higher priority than mining the remaining fossil reserves of coal, oil, and natural gas.

3.5

TABLE 3.6 Global Organic Carbon Distribution

Kvenvoldon, K.A. (1993)
"Gass Hydrogen Geological
perspective and global
change" rev. Geophysics 31,
173-187

Methane Hydrates	10,000 Gigaton
Fossil Fuel	5000 Gigaton
Atmospheric Methane	3.6 Gigaton

changes resulting from widespread methane escape. Unless we do what we must to decrease this incipient danger the probability of harm increases.

BE IT FRIEND OR FOE, IT IS TOO BIG TO IGNORE.

3.6 Worldwide deposits of methane hydrates have been conservatively estimated by the U.S. Geological Survey to contain more than twice as much carbon as all known fossil fuels on Earth. Table 3.6 shows the comparison.[1] Releasing substantial amounts of this greenhouse gas that now rests precariously on the ocean's floor would seriously change Earth's climate.

Although the methane hydrates pose difficult problems, the author has experimented with practical approaches for reducing potential harms by processing these dangerous deposits to produce hydrogen, carbon products, and pure water. Equation 3.1 summarizes the process.

Equation 3.1   Methane Hydrate + HEAT --» Water + Carbon + Hydrogen

Heat required for the endothermic processes of Equation 3.1 can be provided by circulation of warmer water, concentration of solar energy or by harnessing wind, wave, or falling water. In addition to pyrolysis by concentrated solar energy, the author has proven the feasibility of continuous carbon and hydrogen production by an autogenous processes in which as little as 10% of the methane produced is spent to provide the conditions needed for dissociation of the remaining feedstock.

Production of methane from methane hydrates along with organic wastes from sewage disposal operations, landfills, farms, and forests will provide essential supplies of carbon to build the hydrogen powered homes, cars, ships, airplanes, farming equipment, manufacturing systems, and consumer goods of the Solar Hydrogen Civilization. The Solar Hydrogen Civilization will have much stronger, lighter weight, and more corrosion resistant carbon reinforced materials than any previous era.

In solar-rich areas it will be profitable to invest some of the carbon in structures to collect and concentrate the sun's rays to produce the dissociation temperature needed. In wind-rich areas it will be profitable to invest some of the carbon in structures to convert wind and wave energy into electricity to provide for dissociation of methane into hydrogen and carbon. In other areas it will be profitable to utilize some of the hydrogen in autogenous generation of carbon and hydrogen.

The Energy Return on Investment or E-ROI and the Social Return on Investment or SRI are both very high for each of these approaches.

These are important drivers for justifying the scale of activity that is needed to overcome the potential disaster that looms while going far towards solving the energy shortage that confronts Civilization.

These resources must not be ignored nor squandered. The more we convert potentially dangerous methane into carbon products and hydrogen the more successful we will be in achieving our Grand Purpose of Sustainable Prosperity.

Numerous references have recommended collections of carbon dioxide from smoke stacks and then liquefying the carbon dioxide for transporting it to deep ocean trenches for storage. It will be far more desirable to produce carbon by the process of EQUATION 3.1 and to utilize the carbon to build sea-walls, bridges, safer and more efficient vehicles and countless other carbon reinforced structures. This approach does not threaten the oceans with carbonic acid that would form upon mixing carbon dioxide with sea water. Using the hydrogen in fuel cells and/or engines to produce heat and motive power will provide far greater energy-conversion efficiency as carbon dioxide pollution of the atmosphere or ocean is prevented.

DEFENSE OF THE ENVIRONMENT SHOULD BECOME THE PRIMARY OBJECTIVE OF THE DEFENSE AND ENERGY ESTABLISHMENTS

3.7

One of the most important preemptive activities that our defense industry could endeavor is to overcome the threat of methane releases into the atmosphere from methane hydrate deposits. Doing so can be one of the most socially endearing and meaningful activities ever provided by the defense establishment.

Farming and other essential commercial activities along with military readiness are now threatened with impotence because of potential shortages of fossil fuels. Illustratively, if the U.S. armed services could not secure oil from foreign sources and had to fight a war with the same fuel requirements as the Desert Storm War of 1990-91, it could not be a protracted conflict. U.S. strategic oil reserves would only allow U.S. military forces to use the vast assortment of powerful engines, technologies, and weapons that we now depend upon, for about two months!

After two months we would have serious fuel supply problems. Farming and other essential commercial activities would have to increasingly give up oil destined to support food, fiber, and durable goods production in order for military efforts to continue.

We must consider the Earth's vast oceanic methane hydrate deposits as strategic energy supplies and sources of carbon to build the structural wonders of the future. This urgent opportunity to convert an incipient problem into economic progress and environmental betterment deserves a great share of future defense budgets, and most of the subsidies now harmfully supporting depletion of safer continental deposits of coal, oil, natural gas, and nuclear fuels.

The ocean's strange oceanic deposits of methane ice and silt must be converted into carbon products and hydrogen energy. Applying our defense establishment's organizational skills and the experience of the energy industry will help enable conversion from fossil hydrocarbons to hydrogen and will greatly improve the global environment, the economy, and our moral integrity.

HYDROGEN IS THE ULTIMATE ALTERNATIVE FUEL

ILLUMINATING GAS ENGINES

Hydrogen fueled the first internal combustion engines and it will fuel the engines of the future

HOW LONG WILL THERE BE CHEAP GASOLINE AND/OR DIESEL FUEL?

CHAPTER 4.0

Oil is refined into the gasoline, diesel fuel, jet fuel, and bunker oils that power virtually all of the transportation engines on Earth. During the development of the Petroleum Age, the U.S. led the world in oil and natural gas production. U.S. oil production steadily increased from a few barrels in 1859 to about 9,637 million barrels per day in 1970 and has declined gradually to the present point of about 5.853 million barrels per day.

The U.S. leads non-OPEC petroleum production with combined oil, natural gas and natural gas liquids equivalent to about 8.09 million barrels of oil per day. Russia is second in non-OPEC petroleum production with the equivalent of about 6.22 million barrels per day. Figure 4.1 shows oil production by these Super Powers along with other petroleum production rates that are now equivalent to about 75 million barrels of oil per day. OPEC provides about 40% of the world's petroleum production including oil, natural gas, and natural gas liquids.

TABLE 4.1	GLOBAL DAILY PETROLEUM PRODUCTION
USA	8.09 million barrels per day
Russia	6.22
Norway	3.48
Mexico	3.33
United Kingdom	.94
Canada	2.68
Brazil	1.42
Other Non-OPEC	14.19
OPEC	29.46
January 2001 TOTAL	75.00 million barrels per day

World Crude Oil reserves as of January 2001 are shown in Table 4.2 As noted, the U.S. has between 2 and 3% of the world's crude oil reserves.

What is obvious is that the Super Powers, United States and Russia, will not be able to continue to lead the non-OPEC countries in oil production because the combined holdings of these "Super

Producers" amount to less than 7% of the world total reserves. All of the non-OPEC countries combined hold less than 21% of the world's remaining oil reserves. Soon OPEC will control 90% of remaining oil reserves because 60% of the demand for petroleum is supplied by countries that have 21% of remaining reserves.

Soon OPEC will control 90% of the world's remaining oil.

Oil company representatives generally insist that the world has plenty of oil and there is no reason to change to another fuel. One conclusion is that the business plans of oil companies welcome the

TABLE 4.2 GLOBAL OIL RESERVES

COUNTRY	OIL RESERVES[1]	PERCENT OF TOTAL
United States	22.0 BILLION BARRELS	2.14%
Russia	48.6	4.73
Norway	9.4	0.91
Mexico	28.3	2.75
United Kingdom	5.0	0.48
Canada	4.7	0.46
Brazil	8.1	0.79
Other Non-OPEC	87.5	8.51
OPEC[2]	814.5	79.22
TOTAL	1,028.1 BILLION BARRELS	100.00

1.*Oil and Gas Journal, January 2001*

2.*Eleven Countries: Algeria, Indonesia, Iran, Iraq, Kuwait, Libya, Nigeria, Qatar, Saudi Arabia, United Arab Emirates, and Venezuela*

Peak Oil escalation of gasoline, diesel, and L.P. fuel prices that follow depletion of oil reserves. These companies have taken record profits as the U.S. continued to burn the largest share of the world's oil production and now they are poised to increase their profits.

TABLE 4.3 OIL IMPORTS TO THE U.S.

Imports of Crude Oil into the United States by Country of Origin, 2001

COUNTRY OF ORIGIN	1000 BARRELS	COUNTRY OF ORIGIN	1000 BARRELS	COUNTRY OF ORIGIN	1000 BARRELS
1 Saudi Arabia	588,075	14 Argentina	21,013	27 Brazil	4,667
2 Mexico	508,715	15 Trinidad and Tobago	18,562	28 Algeria	3,966
3 Canada	494,796	16 Other	15,874	29 Peru	2,524
4 Venezuela	471,243	17 Indonesia	14,759	30 Thailand	1,751
5 Nigeria	307,173	18 Congo (Brazzaville)	14,430	31 Ivory Coast	1,517
6 Iraq	289,998	19 Australia	12,567	32 Cameroon	1,255
7 Angola	117,254	20 Yemen	8,702	33 Congo (Kinshasa) *	345
8 Norway	102,724	21 Brunei	8,174	34 Qatar	69
9 Colombia	94,844	22 United Arab Emirates	7,802	Total	3,404,894
10 United Kingdom	89,142	23 Oman	7,138	Non OPEC	1,635,274
11 Kuwait	86,535	24 Guatemala	6,485	Arab OPEC	976,445
12 Gabon	51,065	25 Malaysia	5,643	Other OPEC	793,175
13 Ecuador	41,403	26 China,	4,684		

If the world economy is to follow the course of expansion it is imperative to develop renewable energy resources. Petroleum production cannot meet growing world demands for clean energy. Rising prices have followed growing demands, as have profit margins. In order to receive these profits, the U.S. military must be maintained as the world's enforcer of oil shipments to the U. S. and others who join the U.S. in the war against terrorism. Figure 4.1 shows the rise and fall of oil and natural gas production compared to other energy sources.

FIGURE 4.1

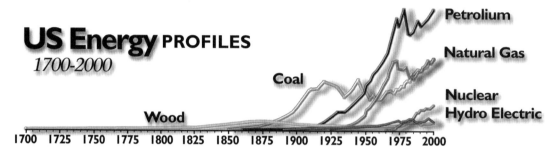

Ref: www.flexible energy.com

FIGURE 4.2

Figure 4.2 shows the expected points that demanded oil production from various sources starts to lag behind production capacities. What is very apparent is the requirement to spend more than ever before to secure oil deliveries to the U.S. These expenditures must be made to build the new infrastructures needed to produce and transport oil from remote places and to provide military protection for these efforts.

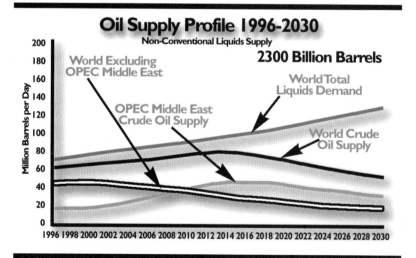

THE U.S. ENERGY ECONOMY:

In order to appreciate why hydrogen is the ultimate alternative, it is important to review the history of energy supplies for the U.S. economy which exceeds all other users of oil. Before the Petroleum Age began with the production of oil in 1859 at Titusville Pennsylvania, energy came from renewable sources such as wind energy and combustion of wood. This quickly changed after oil was discovered and the U.S. and the rest of the world became dependent upon fossil fuels such as coal, oil, and natural gas. Figure 4.1 shows the U.S. energy market and types of energy from 1700 to 2000.

Today, fossil fuels directly supply 80% of the U.S. energy and a substantial portion of the 8% that is attributed to nuclear fuel. This is because nuclear energy is actually produced by very large expenditures of oil, coal, and natural gas. Fossil energy is also used to cut and haul firewood. Hydroelectric plants are also built by work of fossil fueled engines.

From the 1700's to about 1958 the U.S. produced as much or more energy than the domestic market consumed. Through the 1960's, U.S. oil production and the amount of imported oil significantly grew and by 1973 imports of oil amounted to about 35% of U.S. oil consumption. This trend led to present dependence on imported oil for about 60% of the demand. The U.S. now requires more than 11.9 million barrels per day to be brought - by pipelines from Canada or Mexico, or by super tankers to meet the burn rate of 21 million barrels per day.

4.1

TABLE 4.4

OIL LOST FROM OCEAN TANKER DISASTERS

MAGNITUDE*	TANKER	YEAR	LOCATION
257,000	Atlantic Empress	1979	West Indies
239,000	Castillo de Believer	1983	South Africa
221,000	Amoco Cadiz	1978	France
132,000	Odyssey	1988	Mid Atlantic
124,000	Torrey Canyon	1967	UK, Channel
123,000	Sea Star	1972	Gulf of Oman
101,000	Hawaiian Patriot	1977	Hawaiian Islands
95,000	Independenta	1979	Turkey
91,000	Urquiola	1976	Spain
85,000	Braer	1993	Shetland Islands
82,000	Irene's Serenade	1988	Greece
80,000	Aegean Sea	1992	Spain
77,000	Prestige	2002	Spain
76,000	Khark 5	1989	Morocco
68,000	Nova	1985	Gulf of Iran
62,000	Wafira	1971	South Africa
58,000	Epic Colocotronis	1975	West Indies
57,000	Sinclair Petrolore	1960	Brazil
41,000	Burniah Agate	1979	U.S.A-/Mexico
36,000	Exxon Valdez	1989	U.S.A./Alaska
36,000	Corinthos	1975	U.S./Delaware
29,000	Texaco Oklahoma	1971	U.S. East Coast
21,000	Ocean Eagle	1968	Puerto Rico

*Tons

OIL SPILLS:

Oceanic oil shipments are at risk from many sources. Weather hazards, collisions with other ships, running aground; terrorist attacks, and other mishaps confront these giant tankers. Oil spilled instead of being safely delivered causes expensive environmental problems. Figure 4.4 lists some of the largest spills by ocean tankers. How long will it be until the next oil spill causes another high social cost of dealing with the environmental degradation and clean up that follows?

4.2

Hydrogen Fueled Engines

French engineer Jean Joseph Etienne Lenoir invented the first practical internal combustion engine in 1859. Lenoir's engine utilized illuminating gas (a pipeline-distributed mixture of hydrogen and various impurities such as carbon monoxide, nitrogen, and methane) as its fuel. Lenoir's engine had very low emissions. Since then, designers have championed the wonders of internal-combustion piston engines.

German inventor Nickolaus August Otto patented a two-stroke engine in 1861 that ran on illuminating gas and won the *significant discovery* gold metal at the 1867 World's Fair in Paris. Also in 1861 another inventor Aphonse Beau de Rochas, a French engineer, patented the four-stroke engine cycle anticipating that it would be hydrogen fueled. Otto designed an improved internal combustion engine in 1876 that operated on the four-stroke cycle. Otto formed a company that manufactured the first internal combustion four-stroke engines and it continues today mostly making air-cooled diesel engines.

Internal combustion engines evolved over a period of a century into two types. The more common type wastes more fuel, but is cheaper to produce. The more expensive type is far more fuel-efficient. Inexpensive homogeneous-charge, spark-ignition piston engines that were adapted to burn gasoline became the most popular.

The less popular type, "Diesels" utilize stratified-charge, compression-ignition. Air is first compressed sufficiently to be heated to about 700°F or higher and as the piston approaches top dead center, fuel called *diesel fuel* is suddenly injected through very small orifices to be sheared into a spray of fine droplets that quickly ignite in the hot compressed air. Diesel fuel is specially prepared to ignite at a lower temperature in highly compressed air than spark-ignitable fuels such as gasoline, methane, fuel alcohols, and hydrogen.

PETROLEUM-AGE ENGINES

4.3 Lighter-weight, homogeneous-charge engines designed to meet replacement opportunity specifications (by wearing out) have long been favored by car manufacturers and far outnumber Diesel engines. Diesel engines have distinguished themselves by being built for longer life and much more fuel-efficient operation with lower overall emissions per horsepower hour, but have embarrassed themselves by belching smoke including ultra-fine particles at start-up and during low-speed lugging. Diesels are at their worst in congested cities where smelly fumes and smoke pollute the environment during start and stop operation. The heavier Diesels struggle to keep up with traffic as they clatter to move more ton-miles with less fuel. Diesel busses are detested for filling intersections and sidewalks with smoke and odorous fumes. However, higher fuel efficiency provided by stratified-charge operation of Diesel engines justifies the more expensive engine that is heavier and lasts longer than more popular homogeneous-charge engines.

Homogeneous-charge engines utilize a variable valve in the air intake manifold to restrict or "throttle" air that enters the engine. This produces a variable vacuum in the intake

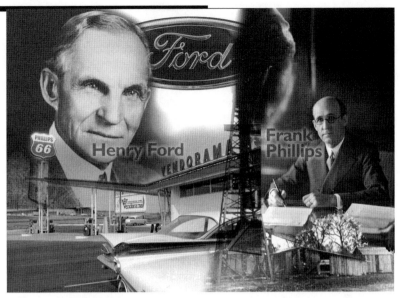

Henry Ford

Frank Phillips

manifold. Transportation vehicles use a foot-feed lever to control the air-throttling valve. Gasoline is metered into the vacuum of throttled air that passes into the engine's combustion chambers and hopefully produces a thoroughly mixed, homogeneous-charge that is spark ignitable in all portions.

Transportation engines were designed to "run rich" with more gasoline than could be burned in the amount of air allowed into the combustion chambers. Additional gasoline cooled the combustion chambers. Residual carbon and lead deposits cushioned the valves. This practice along with the prevailing use of leaded fuel allowed very high profit margins on vehicles with low-cost engines made of materials that otherwise would be prone to earlier failure due to corrosion and wear.

DEPRESSION PRECEDES SPENDTHRIFT ECONOMY:

Glories of the stock market boom of the 1920's brought widespread optimism and belief that technological progress would always be able to spur the economy and reward exuberant speculation. However, the U.S. Federal Reserve raised interest rates in 1928 and 1929 to discourage rampant stock speculation with borrowed funds. This had little immediate effect on stock prices but a very large negative impact on purchases of durable goods. Production of durable goods fell, unemployment rose, and prices of farm products and manufactured goods fell.

Purchasers followed the lead of speculators and expected to have opportunities to buy commodities and agricultural land at much lower prices if they stopped buying for a while. Banks collapsed and uncertainty in the world monetary system brought doubts about virtually everyone's credit. The Great Depression that gripped the world economy in the 1930s was an order of magnitude larger than other depressions before or after it. Government spending for infrastructure, military supplies, and military training facilities put many Germans, Italians, Japanese, Russians, British and Americans back to work and increased the likelihood of war. Hitler and other dictators were moving to

Oil taken to market by tank trucks, rail cars, and pipelines required less capital investment than the production plants and pipelines for illuminating gas. Vehicles were designed to use gasoline, which required much less expensive storage tanks than the high-pressure storage vessels for compressed gas fuel selections. Automobiles went into mass production by efficient assembly lines and with engines that were dedicated to liquid fuels. As oil interests increasingly controlled the temporary fossil economy, cleaner fuels such as illuminating gas were left out of the plans.

4.4

capture oil and other resources.

After WWII, U.S. oil production continued to set new records to supply gasoline and diesel fuel to record numbers of U.S. vehicles and to support the Marshall plan to rebuild war-torn Europe and MacArthur's plan for revitalization of Japan. Mass production of expendable war materials was transformed into mass production and marketing of low-cost, ready-to-use, paper and plastic supplies. Car parts if not cars were designed for replacement instead of repair. Garbage disposal operations encountered exponential increases of paper and plastic products along with refuse from new marketing pressures that kept doubling the volumes of disposable goods such as fast food packaging and diapers. Executives, business-school professors, and stock brokers agreeably clucked "it ain't broke, don't fix it."

Environmental protection rules were enacted to stop most of the industrial incineration and household burning of garbage and other combustibles. Garbage disposal required much larger and more mechanized systems. More petroleum was produced by the U.S. and U.S.S.R. to fuel the economic growth that was struggling to support an expensive cold war between capitalism and communism.

Planned obsolesce and biodegradable themes were sold to the public as a throwaway economy developed. It looked like the public would have to learn to like internal combustion engine exhaust with enough carbon monoxide to kill a person in a matter of minutes, enough lead to dull his children, and enough carcinogens to eventually give everyone cancer.

By 100 B.C., long before the advent of tetraethyl lead, Greek physicians had described lead poisoning. Lead poisoning of children causes hypertension, development problems, and intelligence loss. A prevalent source of lead in the global atmosphere is a metal-organic compound called tetraethyl lead (TEL). General Motors pioneered development of TEL in the 1920's and TEL has eventually sourced 90% of the world's lead pollution. TEL is marketed as an additive to prevent the least expensive gasoline from causing excessive valve-seat recession and piston knock in inexpensive engines.

Cleaner burning fuels with "no-knock" tendencies were well known since the time of Lenoir and Otto's engine developments in the 1860's (including hydrogen, methane, fuel alcohols) but General Motors, DuPont, and Standard Oil united against the use of these less expensive fuels while they created a giant market for leaded gasoline. Hardened valve seats were available in the 1920's to provide very long valve life and were utilized in Diesel engines. Leaded gasoline still supplies 94% of the fuel requirements for vehicles in the Middle East, 93% of the needs of African motorists, 35% of the market in South America, and over 30% of the vehicle fuel in Asia.

After acquiring catalytic reactor technology, General

Motors suddenly lobbied for U.S. phaseout of lead in gasoline, which began in 1975 and was nearly accomplished by the end of 1986. During this phaseout, lead in U.S. human blood samples declined about 78%. U.K. efforts to prevent lead from accumulating in human tissues began in the late 1980's and blood-lead levels have declined about 66% in the last decade. These successes encouraged the European Union to ban leaded gasoline beginning in 2000. There are three remaining large-scale sources of TEL, which are located in Britain, Germany and Russia. Octel, a company that supplies much of the world market with TEL, reported in an SEC filing that between 1985 and 1995 it was able to offset the loss of the large U.S. market by higher pricing and increased sales to developing countries. (India obligingly increased the allowable TEL from 0.22 to 0.56 grams/liter!)

Vehicle pollution is increasing in all populated areas. Developing economies are rapidly increasing the amount of air and water pollution from vehicle exhausts. In Nigeria, roadside lead contamination of 6,000 PPM has been measured. Children may suffer hearing loss, reduced attention span, and learning disabilities if exposed to 600-PPM lead levels. In 1971, vehicle pollution accounted for only 24% of India's environmental degradation. India's motor vehicles now produce about 67% of the pollution in this highly populated country that will double the number of vehicles in a decade. Before recent reductions in TEL content, vehicles in Mexico City polluted the air and water with daily emissions of some 32 tons of lead along with enormous quantities of carbon monoxide and unburned hydrocarbons. Foreign diplomats stationed in Mexico City were provided with "fresh-air retreats" to Los Angeles!

THE L.A. PROTEST:

Los Angeles, California led the world in admitting that it had bad air and became emotional enough about it to seriously demand remedies from automakers. In the 1950's persons living in Los Angeles complained of irritation due to the poor air quality that was noticeably associated with the tremendous amount of fossil fuel demanded by what had already become one of the most productive manufacturing and commercial economies in the world. By the 1970's Southern California's population had burgeoned, freeways made it possible for urban sprawl to quickly enrich developers, and the number of cars and miles traveled each day staggered the imagination. Air quality seriously deteriorated.

4.5

Los Angeles and then many other cities called for legislation to bring relief from mobile and stationary sources of pollution. Federal legislation followed and the Environmental Protection Agency (EPA) was formed in 1970 to curb emissions of air and water pollutants. Automakers testified that the only practical way to deal with emissions from internal combustion engines was to correct the combustion deficiencies with external air pumps that would add air to the exhaust from air-starved engines to allow catalytic reactors to oxidize more of the unburned fuel and carbon monoxide that engines using gasoline produced. Electronic computers would intercede to interpret actions by the human operator and actually control the ignition timing, air/fuel ratio, transmission shift points, etc., to hopefully reduce emissions. In other words, engines would have computers added to intentionally waste fuel, produce hydrocarbons and carbon monoxide emissions, and hopefully these problems would be reduced after leaving the engine by expensive arrangements for further oxidation in the catalytic reactor.

LESS POLLUTION PER CAR,
BUT MORE CARS, MORE PROFITS,
and more pollution. Even the water is polluted!

4.6 But of course the public would have to pay considerably more for a vehicle and it would cost much more for repair parts and maintenance. Oil companies apparently felt left out of the windfall profits that the auto companies were getting for efforts to curb emissions. Soon legislation to add oxygen to the fuel in the form of "oxygenated compounds" or "oxygenates" appeared in California and other states. An alcohol such as methanol CH_3OH or ethanol C_2H_5OH would be added to gasoline so the air-starved engine would get oxygen brought by the new liquid fuel blend that would be sold at a higher price.

Then it progressed to an additive that could be more easily monopolized as tetra ethyl lead had been. A new oxygenate was introduced through the efforts of powerful lobbyists and oil-backed influence groups. Methyl Tertiary Butyl Ether (MTBE) is colorless ether that is made from methanol and methylpropene by an acid-catalyzed addition reaction. It smells something like turpentine and causes discomfort. One study showed that mice breathing high levels of MTBE got liver cancer. Another accelerated study showed that high levels of MTBE in air caused kidney cancer in rats. U.S. Environmental Protection classifies MTBE as a possible carcinogen.

Methyl Tertiary Butyl Ether Production Soared

By 1993, MTBE production had rapidly increased to rank second among all organic chemicals manufactured in the U.S. This volatile organic compound readily enters the atmosphere and mixes in all proportions with water. MTBE prefers to be in water 42 times more than it does air. The Henry's Law Constant (0.024) indicates that it greatly prefers water to air and any vapor that contacts water will deposit MTBE with the water.

Complaints of headaches, dizziness, eye irritation, burning of the nose and throat, disorientation, and nausea have been traced to the presence of environmental pollution by MTBE in Alaska, Montana, Wisconsin, New Jersey and California.

Studies show that the concentration of MTBE in the air is

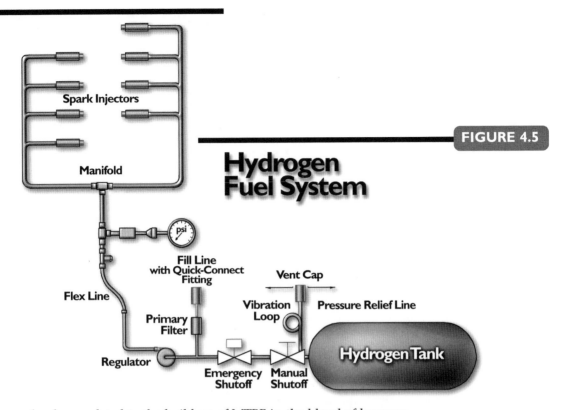

FIGURE 4.5

Hydrogen Fuel System

Spark Injectors

Manifold

psi

Fill Line with Quick-Connect Fitting

Vent Cap

Flex Line

Vibration Loop

Pressure Relief Line

Primary Filter

Regulator

Emergency Shutoff

Manual Shutoff

Hydrogen Tank

closely correlated to the build up of MTBE in the blood of humans and probably all forms of life because of its affinity for water. Mechanics, service station attendants, and commuters on busy streets collect MTBE faster than less-exposed persons. MTBE was detectable in the blood of all persons tested two months after the use of MTBE was suspended in Fairbanks Alaska. Methyl Tertiary-Butyl Ether has been subsequently found as an objectionable contaminant in virtually all ground water in America because it readily travels through the air, diffuses through soils to rapidly invade every moist material, and stays in solution in water where it is very difficult to remove.

HYDROGEN FACILITATES A BETTER "OXYGENATION" SOLUTION
But what about those "anti-tampering" laws?

The author who had preferably used hydrogen as a fuel since the 1960's encountered no-tampering laws that were enacted to enforce the auto and oil companies' expensive and inefficient answers to vehicle emissions. Vehicles that were converted to operation on hydrogen worked much better without the smog control systems. In fact after removing the parasitic pump for delivering air to the exhaust manifold, removing the catalytic reactor, and by-passing the computer for controlling the ignition timing, and blocking the throttle valve wide open, the engine could run better on stratified-charge hydrogen. Tests showed that engines converted to hydrogen operation actually cleaned the air that entered the engine. Figure 4.5 shows the preferred kit of components for conversion of existing engines to

4.7

hydrogen operation.

With a wide-open throttle, more air always enters the engine's combustion chambers. Hydrogen combustion in excess air facilitates very cheap "oxygenation" of the combustion chamber. More air brings more oxygen. Only the hydrogen is restricted by the driver to control the converted engine. Hydrogen combusts in a much wider range of air-fuel ratios than any other fuel. Hydrogen burns well in mixtures ranging from 5 to 70 percent hydrogen in air. With hydrogen there is no need to restrict airflow as required to carefully control air-gasoline mixtures that are only spark ignitable in a narrow band of air-fuel ratios. Figure 4.6 compares poisonous MTBE with air, the "natural oxygenate."

FIGURE 4.6

Poisonous Oxygenate		Natural Oxygenate	
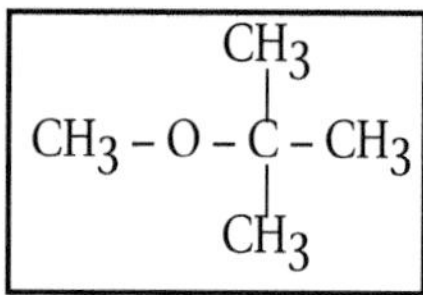 $CH_3 - O - \underset{\underset{CH_3}{\mid}}{\overset{\overset{CH_3}{\mid}}{C}} - CH_3$	MTBE	"AIR"	78% N_2, 21% O_2

MORE AIR, MORE OXYGENATE

4.8 Air is the natural atmosphere that nearly all life on Earth depends upon. Air is essential, healthful, invigorating, and it is free. No contamination occurs if air mixes in all proportions with water. U.S. Environmental Protection Agency classifies air as a safe substance unless it is contaminated by an additive or pollutant. Unless it is polluted, no complaints of headaches, dizziness, eye irritation, burning of the nose and throat, disorientation, and nausea have ever been traced to the presence of adequate natural air anywhere on Earth.

Thus I found that the best oxygenate in the combustion chamber was AIR. What a concept! If you want the fuel to burn more completely and faster, add more air. More air, more oxygenate. It makes the engine work better and air does no harm to the environment. No harm to plants, no harm to animals that breathe, no harm to humans, and no harm to water supplies. Burning hydrogen on a stratified-charge basis in surplus air greatly improves combustion efficiency and reduces heat losses through the piston, cylinder walls, valves, and head areas of the combustion chamber. Engines produce more power, last longer, and clean the air by operating with unthrottled air (the natural, healthful, widely distributed, and free-for-the taking oxygenate).

A synergistic result of using surplus air as the oxygenate is that burning hydrogen in air cleans the atmosphere. Thus, surplus air helps clean more air as it improves engine efficiency and power production.

A kit of components designed for the purpose enables any vehicle to be provided with a parallel hydrogen operation system in about the time of a complete tune-up. After installation of the parallel hydrogen system, the vehicle can retain the option of returning to operation on gasoline. Using the system of Figure 4.5, the driver can operate the vehicle on gasoline and have "bad air" days on demand. Or the driver can select hydrogen and have "good air" days by using hydrogen to produce more power and clean the air while extending engine life.

In order to make this hydrogen conversion kit inexpensive, economy-of-scale production of about 100,000 units per year is needed. (Of course this pales in comparison with the 40 million new cars that are produced each year.) However, 100,000 car conversion kits is about the number of kits needed to enable a sufficient population of cars in a major city to noticeably improve the air. One family car converted to hydrogen operation can clean enough air for three homes each day. So convert your car, and be the *smog buster* that your neighborhood needs!

In addition to working wonders in transportation engines, air and hydrogen enable conversion of other engines that conventionally burn hydrocarbon fuels. Engine-generator sets that produce electricity can be converted from gasoline, diesel fuel or natural gas in the time of a tune-up to operation on hydrogen. Unthrottled air entry into these converted engines provides improved volumetric efficiency and higher overall efficiencies are achieved by the stratified-charge combustion regime. Diesel, propane, and natural gas engines are readily converted for longer life, improved power, and air-cleaning operation.

AVOID GREENHOUSE GAS PRODUCTION
Use the carbon for more profitable purposes:

In addition to cleaning the air, your car will avoid emissions of carbon dioxide. One gallon of gasoline produces about 20 pounds of carbon dioxide. This amount of carbon dioxide occupies about 1,220 gallons of air space. In addition to carbon dioxide emissions from vehicles, power plants, manufacturing facilities, and farming activities cause large emissions of this greenhouse gas. Conventional electricity production releases more carbon dioxide than transportation vehicles. Even eating a meal represents processes that cause volumes of carbon dioxide to be produced.

4.9

Illustratively, each glass of milk produced by modern methods comes from cows that are fed very energy-intensive rations. Milking machines, power filters, and refrigeration systems are powered by electricity from mostly coal-fired central power plants. Diesel trucks that bring the refrigerated raw milk on asphalt or calcined concrete highways contribute to large releases of carbon dioxide. By the time the milk goes through the process of homogenization, pasteurization, and packaging and is offered

TABLE 4.5

Shed tests with hydrocarbon contaminated air compare emissions from engine converted to hydrogen with emissions from same engine on gasoline

Ambient Atmosphere Of Test Shed	28 ppm HC (airborne hydrocarbons)	0.00 (carbon monoxide)	1.00 ppm (oxides of nitrogen)
H$_2$ Idle	19 ppm (eliminated of 9 ppm hydrocarbons)	0.00 (no carbon monoxide)	1.00 ppm (added 0.00 ppm oxides of nitrogen)
H$_2$ Full Power	8 ppm (eliminated of 20 ppm hydrocarbons)	0.00 (no carbon monoxide)	1.7 ppm (added 0.7 ppm oxides of nitrogen)
Gasoline Idle	180 ppm (added 152 ppm hydrocarbons)	24,300 ppm (added 24,300 ppm carbon monoxide)	410.00 ppm (added 409 ppm oxides of nitrogen)
Full Power On Gasoline	192 ppm (added 164 ppm hydrocarbons)	8,100 ppm (added 8,100 ppm carbon monoxide)	110 ppm (added 109 ppm oxides of nitrogen)

NOTES: *ppm = parts per million concentration*
both tests without catalytic reactor in exhaust system
gasoline engine operated according to manufacturers specifications
hydrogen engine operated without air throttling

in refrigerated storage at your grocery market, the fossil equivalent of about 2/3 cup of petroleum has been burned per cup of milk that is served. This is before you transport it to your home in your trusty petrol-fueled car.

About 40,000 pounds of carbon dioxide is released each year per person in the U.S. This amount of greenhouse gas is more than 13 times the weight of a car. Modern power plants produce most of this greenhouse gas, about 3.5 to 4 pounds of carbon dioxide per kilowatt-hour along with about 2 kilowatt-hours of heat rejection into the environment per kilowatt-hour of electricity supplied to ratepayers. Petrol fueled vehicles reject about 3 kilowatt-hours of heat into the environment per kilowatt-hour of energy for propulsion.

Each gallon of petrol (oil, gasoline, or diesel fuel) could be processed at the refinery to provide more than five pounds of carbon and a pound of hydrogen. The carbon can be made into durable, high-value goods. Hydrogen can fuel engines that clean the air.

EVERYTHING THAT BREATHES NEEDS AIR-CLEANING ENGINES:

Tests showed that the air leaving my hydrogen-fueled engines was cleaner than the air entering the engines. This good deed was promptly rewarded with a citation for tampering with my own vehicle. Eventually over a period of about a decade of testing that showed cleaner air in the exhaust of hydrogen fueled engines than the air that entered the combustion chambers, Arizona's legislators provided exemptions from tampering regulations. One official realized what we were accomplishing and said, "Please convert more cars, everything that breathes needs air-cleaning engines."

4.10

Arizona subsequently provided the lowest-cost registration for hydrogen-fueled vehicles and exemptions from being inspected at normal emissions test centers. Therefore, we stopped being harassed and cited for not having a catalytic reactor, exhaust recirculation system, air pump to deliver air into the exhaust stream, or spark advance computer to support gasoline. We were provided with a special tag that allows one occupant to drive in the High Occupancy Vehicle fast lanes of the freeways. We were pleased that HOV lanes became "Hydrogen Operated Vehicle" lanes.

HYDROGEN AS AN ECONOMICAL ENERGY CARRIER
the ultimate alternative fuel.

Solar radiation is the actual source of virtually all energy in the atmosphere, oceans, and outer layers of the Earth's surface. All fossil fuels including coal, oil, and natural gas are carriers of stored solar energy that radiated from the sun to the Earth at least 60 million years ago. Civilization has developed excellent technologies for using such fuels as carriers of solar energy. And now we can improve on these technologies and use hydrogen as the solar energy carrier.

4.11

When an energy carrier (fuel) such as a log is burned, the energy released is solar energy that has been stored as chemical potential energy in the compounds of plant tissues. Table 4.7, lists some common fuels and the heat that can be released when one pound is burned completely. Hydrogen is the best fuel when fuel weight needs to be minimized such as in transportation applications.

A vehicle that gets 20 MPG and weighs 4,000 pounds can be improved to achieve over 30 MPG by trimming it to 2,500 pounds. It takes about six pounds of gasoline to provide the same energy as two pounds of hydrogen. Replacing steel with newly developed carbon

TABLE 4.7

FUEL NAME	CHEMICAL FORMULA	COMBUSTION HEAT RELEASE
HYDROGEN	H_2	61,031 BTU/LB
CARBON	C	14,096
METHANE	CH_4	23,890
PROPANE	C_3H_8	21,053
GASOLINE (OCTANE)	C_8H_{18}	20,529

reinforced components saves much more weight and can greatly improve crash safety. The lighter the fuel for a vehicle, the greater the fuel economy and vehicle performance.

Hydrogen can be removed from any of the hydrocarbon compounds listed in figure 4.7. Illustratively, each molecule of methane (the main ingredient of natural gas) is 25 percent hydrogen by mass fraction, or 80 percent by atomic fraction as four hydrogen atoms are combined with one carbon atom. Although the majority of the atoms in common hydrocarbons are hydrogen atoms, carbon has an atomic mass of 12 compared

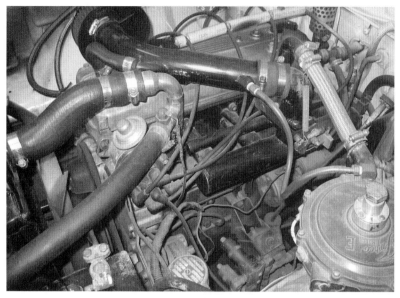
Converted 1978 Dodge engine that cleans the air.

to the atomic mass of hydrogen which is 1. Thus 75 percent of the mass of a molecule of methane is carbon. Seventy-five percent of the weight of methane can be converted into very valuable products that emphasize the spectacular strength, corrosion resistance, and versatility of carbon.

The efficiency of producing fossil fuel substances as carriers of solar energy is very low. Generally, far less than one percent of the solar energy that reaches a green plant is converted by photosynthesis into a fuel value. Farming to facilitate practical photosynthesis consumes considerable energy and further reduces the amount of energy deliverable to the market.

Hydrogen can be produced by several proven technologies that convert solar energy to hydrogen as an energy carrier at more than twenty percent efficiency, including all energy-conversion operations. It will take twenty times less land area to produce hydrogen than to produce the types of compounds from green plants found in fossil fuels or typical fuel alcohols such as methanol and ethanol (CH_3OH and C_2H_5OH).

Removing hydrogen for use as an ultimately clean and lighter energy carrier and using the remaining carbon to produce durable goods makes much more economical use of remaining fossil reserves. Carbon recovered from hydrocarbons can be made into countless durable goods such as wind and wave machines, golf clubs, fuel-efficient cars, and the fastest and most durable aircraft. Hydrogen recovered from hydrocarbons can be used as the "minus-emissions" fuel in engines that clean the air.

Thus we will utilize the best technologies and practices of the Industrial Revolution to efficiently mass produce the carbon materials needed to reinforce solar collectors, wind turbines, wave machines, essential tools, building products, farm equipment and transportation systems needed by every community to achieve sustainable prosperity.

Civilization didn't emerge from the stoneage because we ran out of stones.

THE SUSTAINABLE FUEL
HYDROGEN
AMERICAN HYDROGEN ASSOCIATION

Why change our game plan from burning more oil and getting it by military intervention if need be?

Game theory, as it applies to survival of the fittest, indicates that the most successful individual life forms find ways to cooperate in essential endeavors to achieve commonly beneficial results. This improves the chances of living longer and reproducing more successfully. Civilization evolved as humans cooperated to meet basic needs for food and shelter. Cooperative "civilized" behavior could be extended into virtually every moment of life. City-states developed water distribution systems, irrigated crops, built comfortable structures, and the "fittest" provided care and protection of the young. In many instances the older members of society were honored for knowledge and wisdom that could be taught to improve the lives of succeeding generations. Mentors capable of recording and effectively transferring essential information and technologies accelerated progress.

CHAPTER 5.0

By inventing the technologies needed and cooperating to reach common goals, humans have explored and gained dominion on Earth's lands, oceans, and throughout the atmosphere. In large part, the technologies to do so harness much more energy for much longer periods than a human could possibly apply. Humans cooperated with camels to live in the desert. One relatively short-legged man might have 10 long-legged camels that could cooperatively apply eons of adaptive heredity for desert life to deliver him across otherwise punishing and dreadful deserts.

In more temperate climates, selective breeding and cultivation improved horses, cattle, sheep, dogs, chickens, turkeys, ducks, and numerous species of plants. Farmers stored grain and hay and provided shelter and protection of animals that could reproduce more successfully and thus provide mutual benefits.

Eventually fossil-fueled engine powered equipment greatly increased the rate that humans could control expenditures of energy for work and travel. One man might control 100 or even 100,000 horsepower engines. (One man controlled the 150-million horsepower Saturn V rocket.) Civilized behavior has become more complex with the need to answer the growing problems of resource depletion including air, water, fossil fuels and mineable metals.

One of the complicating aspects of human nature is the human propensity to make improvements such as housing, farms, roads, harbors, and various kinds of vehicles and to do so by overuse of natural resources. Scarcity and hardship lead to hostility over control of diminishing resources. Military action has prevailed throughout human history as the means for redistribution of wealth or resources that can produce wealth.

Exploitation of fossil energy and in turn fossil-age water and other depletable resources allowed humans to rapidly produce a population that is six times larger than could be maintained at any time before coal, oil, and natural gas were harnessed to facilitate the Industrial Revolution. This large population now depends upon burning the fossil equivalent of 190 million barrels of oil each day.

In the 21st century and forever after, we must invent and develop technologies that facilitate the use of renewable resources. In order to greatly increase wealth expansion potentials and to avoid increasing conflict over diminishing supplies of fossil resources, humans must find ways to cooperate in expeditious adoption of a global plan to exploit energy and other resources from outer space.

With the development of specific technologies to produce hydrogen from renewable resources, store it safely, and use it in critical applications, we have opened the door to unlimited supplies of clean energy.

ENERGY FROM OUTER SPACE WILL ENABLE EVERY COMMUNITY TO ACHIEVE SUSTAINABLE PROSPERITY

Technologies for exploiting energy from the sun include photovoltaics, solar thermal devices, wind machines, wave generators, falling water turbines, and biomass-conversion systems. Wind is plentiful in coastal areas and on the plains. Solar energy is abundant in the deserts. Rain and snow runoff from mountains and uplands gathers into rushing streams. Wave energy is available on large lakes and ocean coasts. These are some of our most available, renewable energy sources. Energy from these sources has long been harnessed to generate electricity in many parts of the world.

5.1

FIGURE 5.1

Energy received from such sources as the sun, wind, and waves is usually unsteady or cyclic. Solar energy needs to be collected at the rate that it is available, stored and used at much larger rates in applications such as powering an automobile or appliance. Consider the solar lawnmower of Figure 5.1. In solar-rich areas the lawnmower can be designed to collect solar energy on the relatively small photovoltaic panels as shown for operating a reversible electrolyzer/fuel cell or charging a nickel metal-hydride battery to power the mower for a shorter time of high-energy work.

Solar Cell

NiMetal Hydride Battery

Motor

Solar energy is collected for a week to provide energy that can run the lawn mower for several hours.

In areas where the sun shines fewer hours or at lower intensity, the photovoltaic panel area may need to be increased but of course the grass probably grows more slowly and if you invest in a larger panel array you can utilize any surplus energy to supply supplemental power to your home or other garden equipment. If left where the sun shines most of the time your solar mower will collect and store enough energy while the grass is growing to cut the grass once or twice each week. It illustrates what can be done to gradually collect renewable energy and convert it into a stable, storable form of energy that can be quickly and conveniently delivered as needed.

The single most viable solution for storing and delivering renewable energy comes in some form of hydrogen. Energy from any of the renewable resources can be readily used to generate hydrogen from plentiful supplies of water and biomass wastes.

Hydrogen is a clean-burning fuel that can be used for almost any application in which other fuels are used today. Compared with electricity, which is extremely difficult and expensive to store, hydrogen has been stored by nature for millions of years as a mixed gas in underground natural gas formations. Technological developments over this century have made it easier to store hydrogen in mass-manufactured systems. And hydrogen can be readily transported as a cryogenic liquid in ocean tankers or as a gas in under-ground pipelines at a fraction of the cost of distributing electricity.

Thus hydrogen answers the compelling needs for accumulation and storage of unsteady supplies of renewable energy. Moreover, hydrogen offers major advantages over fossil fuels such as coal, oil, and natural gas. Global reserves of fossil fuels are limited and not easily renewable; and when fossil fuels are mined and burned, they generate air and water pollution. By contrast, hydrogen can be readily regenerated from renewable resources; and when it is burned, it produces clean water and energy such as electricity and heat. Some of the technologies that facilitate the production, storage, and use of solar hydrogen have interesting histories.

FIREPLACE SERVANT:

In the early 1800's, a Scottish preacher dreamed of a way to have his fireplace be his servant. Reverend Robert Stirling studied the hot-air engines of Sir George Caley and wonderful steam engines of James Watt. Stirling was concerned about the danger of explosive failures by boilers and envisioned a much simpler, smaller, and safer hot-air engine that he could fit in his fireplace.

Stirling reasoned that heat from the fireplace could cause a small inventory of captured air to expand and do work on a piston that could somehow be attached to a saw to cut wood for the fireplace or be attached to a pump to deliver water to his household. After expansion, the air would be cooled by heat exchange to the cooler outside air to make it contract to produce lower pressure and then be more easily returned to be re-heated in the fireplace chamber. Reverend Stirling further reasoned that his engine would be more efficient if the air exchanged heat to an "economizer" that saved the heat for quickly reheating the air as it traveled into the fireplace chamber for final heating. In 1816, Reverend Stirling applied for a patent on his new concept for a hot-air engine with an economizer.

Stirling's "economizer" would later be developed into a "regenerator" to serve the purpose of "parking" heat in a heat exchanger and thermal storage media. This enabled heat to be re-added quickly to gases being transferred into the heated chamber. Stirling reasoned that after doing work, gas on the way to the cooler heat exchanger could be pre-cooled by the economizer and on the way back, the economizer would pre-heat this gas before final heat addition in the heated chamber. Modern Stirling-type engines that utilize regenerators hold the world record for converting solar energy into rotating shaft energy.

WAR PROBLEM STIMULATES SOLAR-STIRLING ENGINE COMBINATION:

FIGURE 5.2

During the American Civil War, the South converted a large steam frigate called the Merrimac to an ironclad warship with sloped sides. Renamed the C.S.S. Virginia, but popularly known as the "Ironclad Merrimac" this new development in fighting ships threatened to end the North's naval blockade of the South's harbors. Chief Engineer W.P. Williamson and Naval Constructor John L. Porter revised the Merrimac and added 800 tons of pig iron to produce a mostly submerged ship that was clad with four-inch thick sloping iron armor plates over twenty four inches of oak and pine plates to below the water line.

These formidable iron-plated shields protected it from cannon balls that could easily disable or sink conventional wooden ships. In the first day of engagement at Hamptons Roads harbor, the C.S.S. Virginia rammed and sank the twenty-four-gun wooden hulled Cumberland and turned to engage the fifty-gun wooden hulled Congress. Cannon balls from the Congress bounced off the sloped armor of the C.S.S. Virginia while returned cannon fire ripped through the Congress which was soon destroyed. Unless checked, the South's powerful ironclad steamship was going to sink the North's blockading warships.

The Monitor and Merrimac engaged in battle in battle on March 9, 1862.

About 6 months earlier upon learning about the South's secret plans to convert the steam-powered Merrimac to an ironclad gunship, President Lincoln sought emergency proposals for building a ship that could defeat the Merrimac. About twenty proposals were reviewed and the proposal by John Ericsson seemed to Lincoln to be the best. But Ericsson had a bad reputation with Navy officials. Ericsson had already invented improvements on air compressors, boilers, naval guns and a screw propeller for propelling marine vessels.

Ericsson had designed the screw propeller and built a prototype for the British Navy. But they took his concept to their favorites in the defense industry and when Ericsson complained, they disparaged him in references to the source and refused to compensate him. Then he immigrated to the U.S. and designed a fighting ship called the Princeton with a partner, John S. Stockton. One of the features of the Princeton was an advanced twelve caliber naval gun, the biggest ever tried. His partner, seeking the maximum profit from the Princeton contract cut corners and omitted critical reinforcement of the big gun.

The Tyler administration wanted to be identified with the advancement of American Naval technology. While demonstrating the Princeton's advancements with the Secretary of Navy directing activities, the new gun exploded. Secretary of State Abel P. Ushur and Secretary of Navy Thomas Gilmer were killed. President Tyler escaped death by being sufficiently shielded below deck. Ericsson was disgraced.

Seeking another commission, Ericsson described his new invention of a revolutionary war ship with a revolving turret holding eleven-inch guns to Napoleon III but it was rejected. When the Merrimac development as an iron clad was discovered by Union spies, Ericsson quickly proposed a revolving turret gun ship to defend the Union blockade of wooden ships against the South's iron-clad warship.

Ericsson and military buyers were wary about dealing with each other but a deal was struck. With a $275,000 emergency procurement contract, Ericsson designed and built the mostly-submerged, iron-clad "Monitor" in the Brooklyn Shipyards in 101 days. About the only target it presented was his new invention of a rotating gun turret. Ericsson specified multiple layers of armor to heavily plate his gunboat. This would provide a considerable safety factor over the plate thickness that his tests found to be adequate for resisting the best cannons of the Union Navy. But would this armor resist the mighty Merrimac?

Ericsson was confident the Monitor would not be sunk by gunfire. Ericsson specified two 11-inch smoothbore Dahlgren guns in a turret that could rotate to aim at targets. Monitor's gunners would alternately load the big cannons and blast the Merrimac with 11-inch projectiles. But because of the previous embarrassment with a big gun designed by Ericsson, the Navy strictly limited the amount of gunpowder in the charge that could be used to propel each projectile. Ericsson protested but the Navy feared another embarrassing catastrophe and severely limited the blast stress on the new guns.

The Monitor was quickly launched just in time to engage the Merrimac in what otherwise would have been its second day of devastation on the Union's wooden warships. In the 3.5-hour pitched sea battle that followed on March 9, 1862, neither warship could sink the other. Cannon balls bounced off as the two ships maneuvered pass after pass to fire as fast as possible at each other at close range. The Monitor with its eleven-inch guns in a turret had seriously damaged and stood-off the much larger and more powerful Merrimac with its ten fixed-direction guns. Wooden hulled warships with sail power were suddenly obsolete. Armor plated, engine-powered ships would take their place. Ericsson complained that the battle would have been won if the Navy had loaded his big guns with enough gunpowder to do the job they were designed to perform.

But other improvements were urgently required. Monitor's crew was virtually disabled by the time the battle with the ironclad Merrimac ended. Ericsson had concentrated on getting an ironclad warship in the water quickly but important features were lacking. It did not have a proper smokestack. Toxic fumes from the propulsion engine made them sick. Analysis of the situation brought mixed emotions. Hapless sailors on the Monitor were more threatened by the noxious fumes from the engine than the hostile cannon fire from the Merrimac.

John Ericsson realized that a Stirling engine would be efficient and safer from boiler explosions than steam engines but how could it be heated without fumes? Smokestacks on ships would require heavy plating for protection or they would be easily disabled by hostile fire. Soon John Ericsson invented a way to heat engines, particularly stationery engines, without fumes. Ericsson combined a parabolic solar mirror and a Stirling engine. The parabolic mirror would be constantly adjusted to follow the sun so it could provide heat at a temperature sufficient to operate an engine without combustion.

POWERFUL SOLAR SERVANTS:

As mentioned, the world's record for converting solar energy to grid-quality electricity is held by a type of equipment known as the "solar dish genset."

5.4

FIGURE 5.3

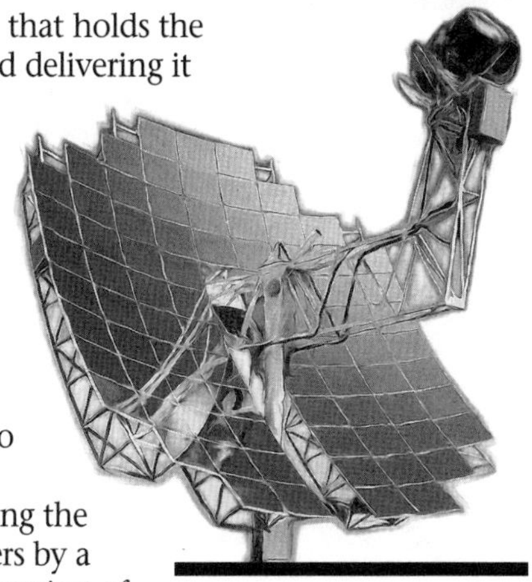

Figure 5.3 shows a modern Solar Dish Genset system that holds the world record for converting solar energy into electricity and delivering it to customers through the electricity grid.

A solar dish genset consists of a parabolic-dish solar concentrator that provides a point focus delivery of intense heat to a Stirling-cycle heat engine that turns an electricity generator. The Stirling-cycle can use gases such as air, nitrogen, or helium but it is most efficient when captive hydrogen is used as the working fluid. In the 1980's, power companies in California (Southern California Edison) and Georgia (Georgia Power Company) used solar dish gensets to deliver over 100 million watt-hours of electricity. These demonstrations established important world records including the most watt-hours of reliable delivery of electricity to customers by a solar-thermal technology and the highest efficiency for conversion of solar energy into electricity for household use.

A reflective parabolic dish collects and concentrates solar energy, which is used to heat hydrogen to about 1,500°F in a heat exchanger of the engine. Heated hydrogen is then moved to another chamber where it does work on a piston and then to an internal heat exchanger called a regenerator where it "stores" heat, and then to another heat exchanger that removes heat at temperatures below 300°F. The heating process expands the hydrogen, while cooling causes the gas to contract. As the hydrogen undergoes cyclical changes in volume, it moves the engine's pistons to do work, and this is conveyed to a generator that produces electricity. Improved efficiency results from the benefit of regeneration or regaining the heat from the internal heat exchanger (Stirling's economizer) where heat is stored as the hydrogen is returned to the first heat exchanger for quickly heating to 1500°F.

The electricity produced by a solar dish genset can be utilized to generate hydrogen by electrolysis of water. By this approach, the overall efficiency of converting solar energy to fuel in the form of hydrogen can exceed 25 percent. In other words, for every 100 energy units (such as joules or BTU) of heat obtained from sunlight, an amount of hydrogen equivalent to 25 energy units (of fuel energy) can be obtained. (Developments presently underway promise considerably higher overall efficiencies.)

By comparison, a generous estimate for the overall efficiency of converting solar energy to the fuel energy of coal (starting from plant life) is less than one percent, perhaps 0.10 percent or less. Thus the production of solar hydrogen can be hundreds of times more efficient - implying that it would take hundreds of times less land area to produce solar hydrogen than to try to generate coal or oil from solar energy. Moreover, the land used to produce solar hydrogen can be an arid desert that couldn't support plant life needed for the generation of crops to replace coal. No fertilizer, no herbicides, and no insecticides are needed for producing solar hydrogen. A new kind of "solar farmer" does the job.

RELEASING THE 'WATER PRODUCER' FROM ABUNDANT SOURCES:

5.5 In the days of alchemy, Theophrastus Paracelsus (1493-1541) noted that when iron reacts with sulfuric acid, "an air arises which bursts forth like the wind." Then in 1700, Nicolas Lemery demonstrated that this invisible "wind" would burn when ignited. Henry Cavendish (1731-1810), experimented with this gas and determined many of its physical properties, called it "inflammable air" and noted

ALTERIVS NON SIT,QVI SVVS ESSE POTEST

AVREOLVS PHILIPPVS THEOPHRASTVS

that burning it produced "nothing but water." But it was Antoine-Laurent Lavoisier who, in 1783, named the element hydro gen - the Greek term "hydro-generator" or "water producer."

Since then, scientists have discovered various ways to produce hydrogen from water or organic matter. These approaches range from the use of microbes or enzymes at ordinary temperatures to reactions that require extreme heat. For instance, blue-green algae can act in the presence of light to split seawater into hydrogen and oxygen gases. This process, known as photolysis, is carried out at ambient temperature. Man-made photolysis can also be performed with selected semiconductors.

The release of hydrogen from biomass, which consists of hydrogen-containing organic compounds, can be accomplished by the action of certain anaerobic bacteria. These microbes produce an enzyme known as hydrogenase, which functions as chemical scissors that can sever hydrogen from complex biomass molecules. Like the algae, many types of bacteria can carry out their activity at ordinary temperatures.

The most common result of anaerobic decay of biomass, however, is the release of a gaseous mixture of methane and carbon dioxide. Methane can be separated from the carbon dioxide by several technologies. This renewable methane can be used to replace natural gas. Doing so provides considerable reduction in greenhouse gas production because the ultimate combustion of the methane only releases as much carbon as was gained by the green plants that used sunlight to convert atmospheric carbon dioxide and water into plant tissues.

Alternatively, hydrogen can be released from water and biomass wastes by addition of heat at temperatures in the range of 1,200-6,000°F (650°C-3316°C). Such elevated temperatures can be readily achieved in a solar-concentrator furnace. Solar furnaces generally achieve high temperatures by delivery of the concentrated rays of solar energy through a point of focus or a small "receiver" hole in the furnace wall. It is common to focus a mirror that is 1,000 to 10,000 times larger than the receiver hole. After entering the receiver hole the solar rays diverge and endlessly reflect off of the walls that define the interior of the furnace. Each reflection results in some radiation being converted into heat. Precisely focused mirrors heat well-insulated furnaces to temperatures that can melt most metals.

Solar furnaces can be built to quickly decompose common substances. Illustratively, at a high temperature water can be dissociated into oxygen and hydrogen. Renewable methane from a biomass digester can readily be dissociated at red-hot temperatures to provide sequestered carbon as hydrogen is released.

Even areas that have relatively little potential for solar energy collection can take advantage of this type of carbon sequestration by thermal dissociation. A relatively small amount of the hydrogen produced by dissociation of methane can be used to cleanly drive the dissociation reaction to release carbon from methane. Similarly, with only a small carbon dioxide production penalty, a portion of the methane feedstock can be utilized to produce the heat needed to dissociate methane and sequester carbon for making durable goods. This provides much greater economic benefit than burning the methane. After accounting the value of carbon for producing durable goods, the hydrogen can be assigned a very low production cost.

Another well-developed process of breaking water into hydrogen and oxygen gases involves the passage of an electrical current through water. This process, commonly referred to as "electrolysis," can be carried out at ambient temperature or at elevated temperatures. Elevating the temperature reduces the voltage required for electrolysis of water. The electrical power needed for such a process can be obtained from falling water, wind, or wave powered electricity generators along with solar energy, using either a solar dish genset or by photovoltaic systems.

Electrolysis can offer the advantage of producing pressurized hydrogen and oxygen without requiring a compressor. By increasing the applied voltage, the electrolyzing process can be made to release the gases at virtually any desired pressure.

Various additional methods involve the use of chemical reagents to release hydrogen from water. If these reactions are performed with the addition of heat, they are called thermochemical reactions; if they involve the use of both heat and electricity, they are referred to as thermoelectrochemical processes. Both types of processes can be designed to regenerate the starting chemical reagents, but the former usually involves more steps than the latter.

An example of a thermochemical process is the use of chlorine and chromium chloride at elevated temperatures to extract hydrogen from steam, through a series of steps. An example of a thermoelectrochemical process is the heating of water (to form steam) in the presence of metallic iron to produce hydrogen and iron oxide (FeO), followed by the recovery of iron from the oxide by the action of an electrical current. The latter operation, being carried out at an elevated temperature, requires less electricity than that needed for ambient-temperature electrolysis of water.

STORING HYDROGEN:

5.6 At ambient temperature and pressure, hydrogen is a gas that is 14 times lighter than air and it is flammable in a wide range of air-fuel ratios. The boiling point (-423.0°F) and freezing point (-434.8°F) are extremely low. These properties have posed difficult challenges for hydrogen storage. Nonetheless, the smallest element can now be safely stored by any of several approaches: compression of the gas; cryogenic liquefaction and/or solidification; adsorption on activated carbon; alloying or compounding with other elements; or various combinations of these technologies.

Since 1898, hydrogen has been shipped in steel cylinders with a capacity of 2,640 cubic inches, holding the gas at a pressure of about 2,000 PSI (pounds per square inch). The author has such cylinders that were first pressure tested and approved for service in 1916 and 1917. These cylinders, painted red to signify hydrogen's flammability, have passed every safety test during more than 85 years of safe service. To ensure safety, such cylinders must meet industry standards, including the use of pressure-relief devices that prevent over-pressurization due to filing errors, mechanical impact, and exposure to fires and other heat sources.

If the storage tanks are reinforced with carbon fiber, they can be made 10 times stronger than regular steel and can store hydrogen at pressures in excess of 20,000 PSI. Such tanks can be mass-produced in lightweight, composite designs that can safely provide 100,000 refill cycles (from empty to full). These tanks can withstand the blast of a full stick of dynamite, an attack with a .357 magnum pistol, and a surface temperature of 1,500°F. The author has used such tanks for 15 years to store hydrogen safely.

Hydrogen can be combined with many metals and selected non-metals to form hydrides or alloys. Certain hydrides at room temperature can accommodate a greater amount of hydrogen than an equivalent volume of liquefied hydrogen. When hydrogen is needed, it can be released from the hydrides by either adding heat or performance of specific reactions.

A popular metal hydride is formed by combining hydrogen with an alloy consisting of 48 percent titanium, 48 percent iron, 3 percent copper, and traces of one or more rare earth metals. An extremely safe storage system can be designed around this type of transition-metal hydride. If the storage vessel for this hydride is penetrated, it will release hydrogen only if heat is added. This provides added leak-control options. Problems with this type of storage include the heavy weight of the host alloy and the necessity of removing heat when hydrogen is added and supplying heat when hydrogen is needed.

A very promising manner of storing hydrogen involves adsorbing it in the capillary passages of what is known as super-activated carbon. This carbon consists of very small particles whose geometry may resemble whiskers or popcorn, and it provides very large surface areas for adsorption. For instance, proprietary materials can provide surface areas in excess of 3,000 square meters per gram. A given volume of these materials can accommodate 2.4 times more hydrogen than an equal volume of compressed gas at 3,000 PSI.

FIGURE 5.4

Even greater hydrogen-to-carbon storage densities may be offered by new forms of carbon, including Bucky balls, whisker scrolls, and nano tubes. Other new proprietary approaches being developed in the author's laboratory may allow hydrogen to be stored at a volumetric density that is on par with that of gasoline. Figure 5.4 illustrates (Figure 3 of U.S. Patent 6,015,064) The American Hydrogen Association has tested the electrolytic production of hydrogen at sufficient pressure to directly load such dense energy storage systems. These are typical technologies that are ready for transfer to new business ventures that will be launched by the International Renewable Resources Institute that is described in subsequent chapters and the Appendix.

ENGINES THAT CLEAN THE AIR:

Virtually any engine, including existing piston engines in applications ranging from mopeds to advanced aircraft, can be converted to operation on solar hydrogen. Engines converted to hydrogen operation are significantly more efficient, and can simultaneously clean the air - that is, the exhaust is cleaner than the intake air. The latter effect, dubbed "minus-emissions," stands in marked contrast to the large quantities of pollutants produced by using petrol to fuel an engine.

5.7

The author taught engineering students at Arizona State University how to achieve the minus-emissions effect using a four-horsepower lawnmower engine. When fueled with gasoline, the engine spewed out high levels of pollutants including hydrocarbons, carbon monoxide, and nitrogen monoxide. But the same engine, when converted to run on hydrogen, actually reduced the concentration of hydrocarbons in the ambient air, while showing no production of carbon monoxide and possibly a slight increase in the level of nitrogen monoxide.

TABLE 5.1

Ambient Air Test		
29ppm HC	0.00ppm CO	1.0ppm NO
Lawnmower with Unthrottled Air & Metered Hydrogen Operation		
Idle: 18ppm HC	0.00ppm CO	1.0ppm NO
Full Power 6ppm HC	0.00ppm CO	2.0ppm NO
Using Gasoline as Fuel in the Same Engine		
Idle: 190ppm HC	25,000ppm CO	390ppm NO
Full Power 196ppm HC	7,000ppm CO	95ppm NO

Table 5.1 shows the minus-emissions performance. These results were obtained without a catalytic converter in the exhaust system. For the conversion kit they devised, the students won the grand prize at the international "WESTEC" competition sponsored by the Society of Manufacturing Engineers in March 1998.

The value of a hydrogen-powered engine can be further seen in terms of its efficiency. Hydrogen holds the world record for burning faster than fossil fuels and in leaner fuel-to-air ratios. These characteristics, combined with direct injection of hydrogen into the combustion chambers of piston engines, enable new thermal efficiency records to be established for most engine types. (Thermal efficiency is the percent of work obtained compared to the input of a unit of energy.) Moreover, if engines currently running on gasoline are converted to operate on directly injected hydrogen, the improvement in fuel efficiency can be 20-50 percent. (Fuel efficiency is the work accomplished or the distance a given mass is transported by a unit of fuel energy.)

Overcoming Past Problems

Difficulties encountered with conventional approaches for introducing hydrogen include loss of power because of reduced ability to intake a sufficient air-fuel mixture and backfiring into the intake manifold. A better approach starts with open throttle air entry into the combustion chamber. Unrestricted air flows into the combustion chamber during the intake stroke. With more air in the combustion chamber, hydrogen is directly injected in larger amounts when needed to produce more power than with gasoline. It is particularly beneficial to have surplus air in the combustion chambers to insulate the combustion process of "stratified-charge" hydrogen. The surplus air provides expansion and useful power before heat is lost to the piston, cylinder, or head components. This type of stratified-charge combustion is much more fuel efficient than homogeneous-charge operation, and the engine components run cooler with less friction and produce much less sludge and varnishing from degraded oil.

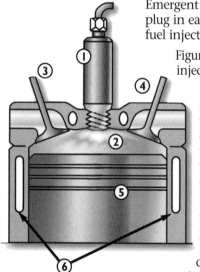

FIGURE 5.5

(1) Spark-Injector
(2) Stratified charge heats the air in the chamber before it heats the cylinder walls
(3) Intake Valve
(4) Exhaust Valve
(5) Piston
(6) Coolant Passages

To optimize the conversion of almost any engine to run on hydrogen, a new device called a Spark-Injector Plug (developed by Emergent Corporation) can be used. This device replaces the spark plug in each cylinder of the engine and combines the functions of fuel injection and spark ignition.

Figure 5.5 shows the Spark-Injector device that combines fuel injection and spark ignition.

Smart Spark-Injector Plugs enable hydrogen to be burned in excess air to produce steam at 4,000°F or higher, if desired, to start very efficient expansion of the gases in a combustion chamber. Heat released from the fast-burning hydrogen is utilized to expand the excess air and to accomplish useful work, such as propelling a motor vehicle or producing electricity.

It should be noted that in the case of gasoline engines that burn homogeneous mixtures of fuel and air, the gases produced in the combustion chamber are often over 5,000°F. Much of this heat is promptly transferred to the piston, cylinder walls, and components of the head, leading to wear and tear on the engine parts and degradation of the lubricating oil. On the other hand, the combustion of hydrogen at 4,000°F is preferably stratified within (enveloped by) excess air, which insulates the combustion chamber surfaces, reducing wear and protecting the lubricant film. Thus, if hydrogen is used under these optimized conditions, the engines run cooler, last longer, and can deliver greater power.

Currently several companies including Larsen Radax Corporation of Chandler, Arizona, are developing new engines to take advantage of these hydrogen combustion characteristics. Early tests indicate that the engine being developed by Larsen could set a new world record for thermal efficiency while virtually eliminating oxides of nitrogen in critical areas of application.

MULTITASKING AIR-CLEANING ENGINES
TO DOUBLE ENERGY-UTILIZATION EFFICIENCY:

Multitasking is an age-old procedure for improving utilization efficiency. Horses were multitasked to pull farm implements and to take farmers to town in wagons or buggies. Dogs were often multitasked to be great friends and confidants, herd flocks of sheep, hunt game, and in snow covered areas to pull dog sleds.

5.8

Some automobiles are multitasked for driver education and sex education.

Energy utilization efficiency could be doubled by utilizing automotive engines to power electricity generators. Collecting the heat not converted into electricity for applications ranging from crop drying to home air-conditioning enables overall energy-utilization efficiency to be doubled. One way is to rebuild automobile engines with Spark-Injector systems and thermochemical regeneration units and place them in service as essential components of cogeneration packages as shown in Figure 5.6.

Another way is to utilize the automobile as a participant in a multitasking cogeneration regime. At night or during other times that the vehicle is not on the road it is hooked up to the electricity grid, a natural gas line, and to the domestic water supply for the dwelling. The engine powers the vehicle's reversible generator to produce electricity for local needs and for deliveries through the grid to customers who automatically buy electronically bid and metered electricity. Heat not converted into electricity is used for cogeneration purposes as illustrated in Figure 5.6.

FIGURE 5.6

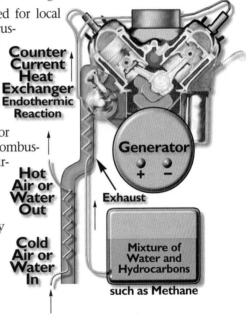

In both approaches any electricity not required for local purposes or for delivery through the grid to other customers is utilized to make hydrogen. In this mode of operation, low-cost methane from landfills, sewage and farm waste disposal operations is converted into high quality hydrogen and utilized by a reversible electrolyzer that may be stationary or on-board the vehicle. Efficient pipeline delivery of low-cost fuels provides production of hydrogen for hybrid vehicles or hybrid homes and businesses that combine internal combustion engines and fuel cells to further extend the air-cleaning and energy efficiency benefits offered.

In a recent summer period of two weeks, Boston area customers paid an extra $80 million for electricity because of 800 MW of lower cost electricity from Maine could not be delivered during peak demand times on the overused grid. Similar grid constraint problems occurred in California, Pennsylvania, Kansas, Missouri, and West Virginia. The estimated cost for these electricity grid constraints during the summers of 2000 and 2001 exceeded $1.8 billion. In addition, California customers suffered much larger penalties due to manipulation of electricity bids as utilities blamed deregulation for soaring wholesale prices. As energy demands grow these problems will magnify.

$$\Delta + C_xH_y + xH_2O \rightarrow$$

$$xCO + (1/2\ Y+X)\ H_2$$

EQUATION 5.0

Utilization of automotive engines in cogeneration systems will avoid the grid capacity problem by production of electricity where it is needed as high quality engine heat is cascaded into manufacturing processes, crop drying, canning needs, water heating, clothes drying, cooking and other applications to further reduce overall energy consumption and grid demand.

These cogeneration regimes could be rapidly achieved by systematic conversion of millions of existing engines to the purposes illustrated in Figure 5.6. Engines that clean the air as they operate in cogeneration systems can double energy-utilization efficiency and securely provide U.S. requirements for energy. Present U.S. demand for about 100 Quads of energy could be reduced to less than 60 Quads of energy as the air is cleaned by converted engines that provide a much larger return on investment in their new multitask applications.

Much of the methane for pipeline distribution can be provided by farmers and other entrepreneurs who convert organic wastes into a new cash crop after supplying their own needs for electricity and heat. Municipalities can do the same to produce revenues from what is now largely wasted by inefficient disposal of sewage and garbage.

Imagine a candidate for mayor who says "Vote for me and I will institute a program to collect your wastes without taxes and use the renewable hydrogen and/or methane that we extract to operate our municipal fleets without incurring any further costs for diesel fuel and gasoline. Following this we will offer bond financing of industrial parks that will provide many new jobs as we take the carbon out of the methane and produce durable goods. Engines in city vehicles and other cogeneration applications will use hydrogen and clean the air… and we will breathe easier knowing that we have done our fair share to improve our health as we improve energy security and reduce costs."

ELECTROCHEMICAL CELLS MAKE AND USE HYDROGEN:

5.9 In 1800 Volta constructed a device that would produce a large flow of electricity. Volta invented an "electric battery"- the first in history. Initially, he used bowls of salt solution that were connected by arched bridges of metal dipping from one bowl into the next, one end of the arc being copper and the other tin or zinc. Volta made the battery more compact by assembling small round plates of copper and zinc, plus discs of cardboard moistened in salt solution. He stacked copper, electrolytic cardboard and zinc in repeating units. A wire attached to the top and bottom of this "Voltaic pile" conducted an electric current when the circuit was closed. This effect would last until the anodes were oxidized or taken into solution by the electrolyte.

Within six weeks after Volta's 1800 publication of voltage production by a battery, two English scientists, William Nicholson and Sir Anthony Carlisle, reported direct current electrolysis of water which was achieved by application of voltage across two metal electrodes in various water solutions. Water was dissociated into two volumes of hydrogen and one volume of oxygen when a current passed through the water from one electrode to another. Electrochemistry was born.

An attractive approach to using hydrogen to generate electricity was provided by a device called a fuel cell. In 1839 Sir William Grove, another English scientist, discovered a way to reverse the electrolysis process and invented the "fuel cell" battery. This device functions through an entirely electrochemical process and avoids combustion of the fuel. It therefore offers an environmentally friendly manner of generating electricity. Moreover, the efficiency of converting the chemical potential energy of hydrogen into electricity can be very high, theoretically about 83 percent at room temperature.

THEORETICAL EFFICIENCY COMPARISON OF IDEALIZED HEAT ENGINES AND FUEL CELLS

5.10 In 1822 Sadi Carnot, an engineer in the French military, reasoned that the maximum possible heat engine efficiency would be equivalent to the highest temperature (T_H) reservoir minus the lowest temperature (T_C) reservoir divided by the highest temperature reservoir.

$$\varepsilon = \frac{T_H - T_L}{T_H} \quad \text{or} \quad 1 - \frac{T_L}{T_H} \qquad \text{Equation 5.1}$$

Grove discovered that platinum electrodes in sulfuric acid could be fed or "fueled" with the gases produced by electrolysis to provide a new kind of battery. Instead of consuming metal plates to produce an electric current, the fuel cell could operate continuously so long as hydrogen was supplied to one electrode and oxygen to the other. Various types of fuel cells have been developed, but they generally operate on the same basic principles. A simple fuel cell consists of two porous electrodes (anode and cathode), with an electrolyte between them. Hydrogen is supplied to the anode, where it is catalytically broken into positively charged hydrogen ions (H+) and negatively charged electrons. The hydrogen ions migrate through the electrolyte toward the cathode, while the electrons are conducted through an external circuit, producing electricity that can be applied to useful purposes. At the same time, oxygen (or air) is introduced at the cathode, where it is converted into negatively charged ions. The hydrogen ions and oxygen ions and/or OH- ions combine to form water (H_2O), which thus becomes a by-product of the process.

Subsequently, Michael Faraday showed that a electrolizer efficiency is the ratio of hydrogen production compared to electricity consumed. Later, Walther Nernst and Willard Gibbs derived the same result by thermodynamic measurements and showed that the maximum possible efficiency of a fuel cell would be limited to the ratio of the chemical free energy change "G" to the thermal heat release "H" of the reaction:

$$\mathcal{E} = \frac{\Delta G}{\Delta H} \qquad \text{Equation 5.2}$$

Thus for the hydrogen-oxygen fuel cell at standard temperature and pressure this maximum efficiency would be limited to:

$$\mathcal{E} = \frac{\Delta G}{\Delta H} = \frac{237 KJ/mole}{286 KJ/mole} = 83\% \qquad \text{Equation 5.3}$$

Compare this limit to the heat engine efficiency limit for operation between the hydrogen combustion temperature of 6000°F and 100°F heat-rejection. Carnot efficiency is based upon absolute temperature, therefore the hydrogen-oxygen

temperature combustion reaction could achieve 6460°R and reject heat at 560°R. The theoretcal efficiency limit for such an ideal Carnot cycle heat engine would be:

$$\mathcal{E} = 1 - \frac{560°R}{6460°R} = 91.3\% \qquad \text{Equation 5.4}$$

Table 5.2 compares the ideal heat engine and ideal fuel cell efficiencies for the same hydrogen-oxygen reaction:

$$H_2 + 0.5\,O_2 \rightarrow H_2O \qquad \text{Equation 5.5}$$

TABLE 5.2

IDEAL HEAT ENGINE	IDEAL FUEL CELL
$H_2 + 1/2\,O_2 \rightarrow H_2O$	$H_2 + 1/2\,O_2 \rightarrow H_2O$
$\mathcal{E} = \dfrac{T_H - T_L}{T_H}$	$\mathcal{E} = \dfrac{\Delta G}{\Delta H}$
$\mathcal{E} = \dfrac{6460 - 560}{6460}$	$\mathcal{E} = \dfrac{237 KJ/Mole}{286 KJ/Mole}$
$\mathcal{E} = 91\%$	$\mathcal{E} = 83\%$

Five major types of fuel cells are making progress to various markets: proton exchange membrane; alkaline; phosphoric acid; molten carbonate; and solid oxide membrane. Each type is under development for a particular market. Variations among them occur mainly in terms of their electrolyte composition and operating temperature. In the first three cases, the electrodes are coated with or in other ways participate with a catalyst such as platinum. The other two operate at higher temperatures and catalysts may not be necessary.

Engine development for major vehicles shows that the market supported inefficient engines and cheap gasoline. While it has long been possible to provide stratified-charge engines that achieve 40% or greater fuel efficiency, the public has bought vehicles with poor fuel economy. Fuel cells that are potentially able to operate at 50% efficiency will more likely be packaged for operation at 30% efficiency because manufacturers opt for the lowest production costs and buyers seem to be willing to pay high life-cycle fuel costs.

Fuel cell stacks for automotive applications require a high power density (electrical output per unit volume of the stack). Ballard Power Systems has been developing fuel cell stacks with increasing power density; which has improved from 3 kilowatts per cubic foot to over 30 kilowatts per cubic foot.

FIGURE 5.7

Proton Exchange Membrane Fuel Cell

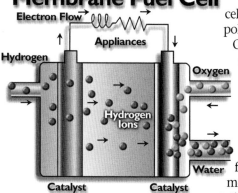

Electron Flow

Appliances

Hydrogen

Oxygen

Hydrogen Ions

Water

Catalyst   Catalyst
Solid Polymer Electrolyzer
Typical Operating Temperature
190 - 230°F (88 - 110°C)
Typical Efficiency 35-60%

The PEM (proton exchange membrane) fuel cell is being developed to replace the lead acid battery and will compete with internal combustion engines in motor vehicles. At the anode, a catalyst converts hydrogen fuel into hydrogen ions (protons) and electrons. The electrons are conducted through an external circuit to provide the electrical current to one or more electric motors that propel the vehicle. The protons migrate through a polymer electrolyte to the cathode, where the catalyst helps them combine with oxygen from the air to form pure water and heat.

Proton Exchange Membrane (PEM):

Operating up to about 210°F(99°C), "PEM" fuel cells are the coolest variety. The electrolyte is a solid polymer membrane that conducts only hydrogen ions. Operating efficiencies are the ratio of electricity produced to the heating value of the fuel consumed. Efficiencies of 30-50 percent are typical where cost and weight savings are important criteria for design. Ballard Power Systems of Burnaby, Canada, is one of the leaders in the development of PEM fuel cells for use in motor vehicles and stationary power plants. Ballard has secured contracts from or set up joint ventures with most major auto manufacturers, including Daimler-Benz, Ford, General Motors, Chrysler, Honda, and Nissan.

Alkaline Fuel Cell:

As the name suggests, an alkaline fuel cell contains an alkaline electrolyte such as potassium hydroxide. Francis Bacon of England began work on this type of fuel cell in 1932, and it has been in routine use on spacecraft since 1965. Today it supplies power aboard NASA's space shuttle. This fuel cell's efficiency can exceed 50 percent, depending upon the output requirements.

Phosphoric Acid:

Operating at a temperature of about 300-400°F (150-200°C), the "phos-acid" or "PA" fuel cell employs phosphoric acid as the electrolyte. This type of fuel cell is targeted for commercial applications, such as hospitals, hotels, and nursing homes. An efficiency of 40 percent or more is typical for such applications. ONSI Corporation and International Fuel Cells (sub-

FIGURE 5.8

Alkaline Fuel Cell

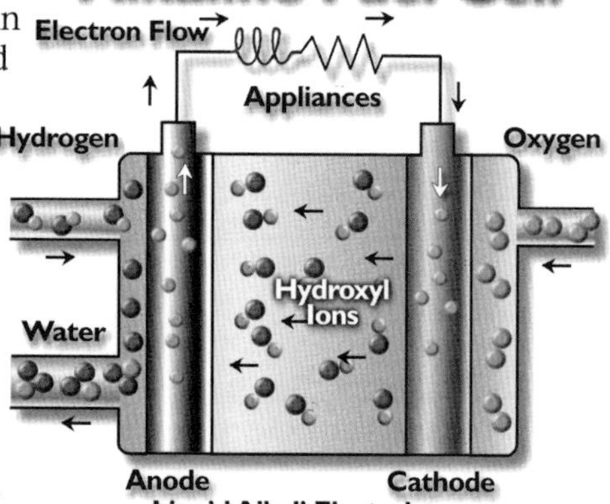

Electron Flow

Appliances

Hydrogen

Oxygen

Hydroxyl Ions

Water

Anode   Cathode
Liquid Alkali Electrolyte
Potassium Hydroxide
Typical Operating Temperature
210 - 480°F (100 - 250°C)
Typical Efficiency 45-70%

FIGURE 5.9

Phosphoric Acid Fuel Cell

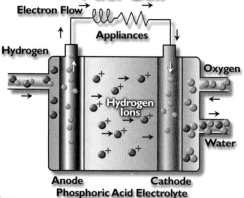

Electron Flow

Appliances

Hydrogen

Oxygen

Hydrogen Ions

Water

Anode   Cathode
Phosphoric Acid Electrolyte
Typical Operating Temerature
300 - 430°F (150 - 220°C)
Typical Efficiency 35-50%

sidiaries of United Technologies Corporation) have developed stationary power plants based on phos-acid fuel cells.

Molten Carbonate:

The electrolyte is a carbonate salt mixture heated into a molten state. At the operating temperature of 1,200°F (650°C), the electrolyte is a good conductor of ions, the electrodes do not need a catalytic coating, and the fuel quality can be lower than for the cooler devices. About 50-60 percent efficiency may be achieved if first cost and weight are not the most important factors, in design. In terms of applications, molten carbonate fuel cells are in competition with phosphoric acid fuel cells. Moreover, they can work well in "cogeneration" systems, as can solid oxide membrane fuel cells.

Solid Oxide Membrane:

These are the hottest fuel cells, operating at about 1,800°F (1000°C). The high temperature provides adequate conductivity of oxygen ions through the electrolyte, which is made of an advanced ceramic material. It is generally expected that these units will be combined with a heat engine such as a gas turbine. Combined cycle efficiencies of over 60 percent seem reasonable for large industrial and utility applications. Figure 5.12 shows a novel application of a TESI engine genset in which some of the exhaust heat is utilized to produce the operating temperature for a fuel cell such as a solid oxide membrane, molten carbonate, phosphoric acid or alkali types.

EFFICIENT COGENERATION:

Central utilities often waste about two-thirds of the heat they generate, by releasing it into the atmosphere at the condensers. An alternate approach that can capture and use this type of heat, provides for the use of a smaller engine driving an electricity generator that can be placed at the site where electricity is needed. This modular engine gen-set and heat exchanger system can recover heat from the engine coolant and exhaust, for use in various heating applications.

Utilization of the thermal as well as the electrical energy generated by an engine is called "total energy" or "cogeneration." It allows for much more efficient use of fuel energy than in central power plants, generally more than doubling the fuel efficiency to operate a home, factory or farm. This combination enables much higher overall efficiency along with an air-cleaning operation.

FIGURE 5.10

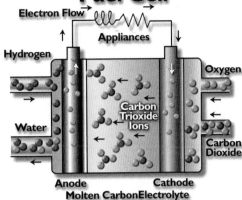

Molten Carbonate Fuel Cell

Electron Flow → →
Appliances
Hydrogen
Oxygen
Carbon Trioxide Ions
Water
Carbon Dioxide
Anode   Cathode
Molten CarbonElectrolyte
Typical Operating Temperature
930 - 1,290°F (500 - 700°C)
Typical Efficiency 40-55%

FIGURE 5.11

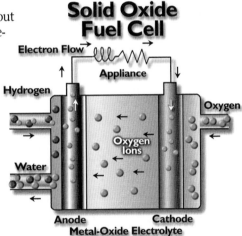

Solid Oxide Fuel Cell

Electron Flow → →
Appliance
Hydrogen
Oxygen
Oxygen Ions
Water
Anode   Cathode
Metal-Oxide Electrolyte
Typical Operating Temperature
1,300-1,830°F (705 - 1000°C)
Typical Efficiency 40-60%

FIGURE 5.12

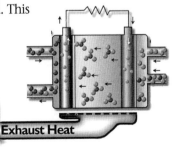

H₂
TESI Unit
Exhaust Heat

THE BIG PICTURE:

5.11 To maintain the energy supplies needed by modern cities, it is desirable to urgently establish what may be termed "renewable energy parks"-sites where the most abundant local renewable resources are converted into electricity and hydrogen. For instance, these parks would be energized by wind in the mountains and plains, waves along the coasts, solar dish gensets in the deserts, and biomass wastes near the cities. These parks will supplant power plants that run on coal, oil, natural gas, and nuclear fuels.

Isolated energy parks, however, will not suffice. We need a "continental-scale" invention that connects the parks and unites the hydrogen energy supply of an entire continent to efficiently meet the needs of cities. Toward this end, the author has made extensive studies that show the suitability of existing natural gas pipelines for transporting renewable methane and/or hydrogen. By using these pipelines, solar hydrogen and/or methane will be delivered throughout the United States, at a fraction of the cost of delivering equal amounts of electrical energy. On the way to market, the infrastructure of pipelines will also deliver hydrogen and/or methane to depleted oil and natural gas wells for storage. (See figure A-1 on page 206)

Depictions of a continental scale system with wave parks, wind parks, solar parks, bio-waste parks and methane hydrate parks sending hydrogen and/or renewable methane to market by natural gas pipelines and storing surplus energy in depleted oil and gas fields of Pennsylvania, Texas, Oklahoma, Louisiana, Kansas, Wyoming, Mexico, and Canada. This continental-scale invention recharges depleted energy storage reservoirs. America can turn back the energy clock to the time when fossil fuels were first discovered.

In addition, the continental-scale invention will incorporate the electricity grid. At off-peak periods, the grid will deliver renewable electricity to areas of heavy intermittent energy use. At these sites, the electricity will drive electrolyzers that provide pressurized supplies of hydrogen and oxygen for storage. During periods of peak energy demand, stored hydrogen will be used in cogeneration engines and fuel cells, without taxing the capacity of the grid for delivery from distant generation sites.

Ultimately, we must think on the global scale. Just as cryogenic tankers currently deliver liquefied natural gas across oceans, it is practical to transport vast amounts of hydrogen from continent to continent. In case of inevitable accidents, spilled hydrogen will quickly escape into the atmosphere, without being the threat to wildlife that oil spills impose. Escaped hydrogen will migrate through the atmosphere or react with oxygen in the air to form water, which will join the natural water cycle.

Of course, the future will come with its own share of problems and challenges. But the problem of satisfying ever-increasing demands for energy can be solved by solar hydrogen.

SUMMARY:

Civilization has important work to do as we launch the Renewable Resources Revolution and wars or rumors of war must not distract this essential endeavor.

One reason that no nation should go to war for oil is because much more beneficial options are available to utilize plentiful renewable energy and carbon that is going to waste. Doing so will provide hydrogen for allowing existing engines to clean the air, last longer, and produce more power. Durable goods made of carbon and carbon-reinforced structures will provide stronger, lighter, safer and more durable products. Sustainable economic development far exceeding any temporary war economy is provided. Instead of sacrificing resources, the Renewable Resources Revolution creates energy-intensive products that can provide sustainable wealth expansion. What the world needs now is recruitment of persons in every continent from every age and background who desire the rewards of global wealth expansion instead of the pain, misery, and waste of war.

GOING UP!

Dreams of flight must be as old as man's contemplations of birds that winged their way through the air and the drifting of clouds that sailed across the sky. Birds fly where they will and see everything! But how would man support his weight after entering the thin air?

Since the observations of Archimedes (287 BC – 212 B.C.) regarding buoyancy as a function of the weight of water displaced, innovations of lighter-than-air concepts for lifting bodies were waiting to be tried. Who would be the hydrogen heroes?

Hot air balloons were tested by two brothers, Etienne and Joseph Montgolfier. In 1782, Montgolfiers watched smoke rise into the sky, and reasoned that the force lifting the smoke would also lift something holding the smoke. They tested their theory and watched paper bags filled with smoke rise into the air. On September 19, 1783, in Versailles, the Montgolfier brothers flew the first passengers in a basket suspended below

CHAPTER 6.0

a hot-air balloon. They showed a crowd of about 130,000 how a sheep, a rooster, and a duck could travel through the air. The eight-minute flight carried the distinguished barnyard passengers two miles from the lift off.

On October 15, 1783, passenger service for humans started. The Montgolfier's new hot air balloon rose 84 feet with their first human passenger, Jean-Francois de Rozier. A short while later, on November 21, 1783, de Rozier made a free ascent in a balloon and flew from the center of Paris to the suburbs, about 5.5 miles in 25 minutes. On January 19, 1784, a huge hot-air balloon built by the Montgolfiers carried seven passengers to a height of 3,000 feet over the city of Lyons, France.

Numerous problems were encountered with hot-air balloons and on August 27, 1783 the first hydrogen-filled balloon was tested in Paris by Jacques Alexandre Cesar Charles of the French Academy of Science. It was immediately apparent that hydrogen balloons had the advantages of being more buoyant because of the much lower density of hydrogen compared to hot air. Hydrogen does not require dangerous combustion to heat air, and they could stay aloft long after the hot air balloons had used up all the fuel that could be carried.

If air is heated by a fire on the ground the balloon will have buoyancy until the air cools, then the balloon descends. When a fuel is burned in the gondola to re-warm the air, chances of setting the bag on fire increase. Hydrogen constantly produces more than ten times the lifting force without a fire. Charles experimented with silk bags that were filled with hydrogen and learned how to treat the silk with an elastic gum to seal it.

Charles quickly built a larger hydrogen balloon named La Charlière and, on December 1, 1783, he traveled 27 miles to Nesle with his assistant M.N. Robert. A crowd of 400,000 people watched their takeoff. The basket that carried the aeronauts hung from a net that covered the entire balloon and distributed the weight evenly. The bag opened at the bottom and had a valve at the top to release hydrogen when they decided to land. Water or preferably stones served as ballast that could be dropped after proper warnings were shouted! A barometer and a thermometer measured the pressure and temperature of the air and the La Charliere hydrogen balloon provided the first scientific flight data.

First Flight Across the English Channel

On January 7, 1785 Jean-Pierre Francois Blanchard and John Jeffries, an American doctor, made the first airborne trip across the English Channel in a hydrogen balloon. During the 2.5-hour flight, the balloon suddenly lost enough gas to almost fall into the English Channel. In order to stay in the air, the worried aeronauts dumped their ballast, most of their cargo, and almost all of their clothing. However, the shivering but triumphant aeronauts managed to deliver the first international airmail – a package from England to France.

Jean Pierre Blanchard brought his hydrogen balloons to the new United States of America and in January of 1793 made the first free flight in the United States by flying over Philadelphia. Blanchard continued his balloon exhibitions until he was killed in a balloon accident in 1809, but his widow, Madéleine-Sophie Blanchard, continued. She was the first female aeronaut. On July 7, 1819, she flew over Paris but her balloon caught fire from fireworks that she was setting off. She managed to land the balloon on a rooftop but then a gust of wind blew her off the precarious perch. Her neck was broken and she died because of the fall from the rooftop.

Pierre Jullien another Frenchman built and demonstrated a streamlined model airship named Le Précurseur at the Paris Hippodrome. Jullien's model airship had a steering rudder, diving and climbing elevator, and passenger gondola mounted under the balloon. It was indeed a precursor of airship designs for the next two centuries. A light wire frame stiffened by trusses maintained the form of the gasbag. The Le Précurseu departed from the hot air style of spherical lifting body and looked surprisingly like a twentieth-century airship. A clockwork motor drove two airscrews to propel the hydrogen airship

Jullien's model of an airship was carefully studied by Henri Giffard, a French engineer. Giffard built a cigar-shaped, non-rigid bag for containing 113,000 cubic feet of hydrogen. It was 143 feet long with a 3-horsepower (2.2-kilowatt)-steam engine to drive its propeller, and it had a rudimentary vertical rudder. A gondola was suspended from a pole that hung from a net surrounding the balloon. The 250-pound engine, 100-pound boiler, 150-pound water reservoir, and the coke required to fire the boiler required most of the available lifting force.

Giffard flew about 17 miles on September 24, 1852 from Paris to Trappes at about 6 miles per hour. Giffard's underpowered airship could be steered in calm weather but any significant breeze caused his airship to be unmanageable.

Giffard planned to build a much larger balloon to be 1,970 feet long, 98 feet in diameter at the middle, and with a capacity of 7,800,000 cubic feet of lifting gas. It was to be powered by a steam engine that weighed 30 tons to propel the airship at 45 miles per hour. Because of its high cost, this huge airship was never built.

At ambient temperature and pressure hydrogen is over 14 times lighter than air. Heating air to the boiling point of water to reduce its density would still provide ambient-temperature hydrogen with the advantage of being over 11 times lighter. This provided more than 11 times greater lifting force that increased to 14 times as the hot air cooled.

Numerous advantages of hydrogen as a lifting gas brought efforts to produce hydrogen more conveniently. Reacting iron filings with sulfuric acid was found to be reliable. However, precautions had to be taken to safely handle sulfuric acid and the aggressive fumes from the reaction. The highly exothermic reaction of iron and sulfuric acid readily released hydrogen that could be cooled and separated from the droplets of sulfuric acid by more iron filings or an asbestos filter. Equation 6.1 shows the general reaction of iron being oxidized as two electrons are contributed to hydrogen ions to form hydrogen atoms that are released to form a diatomic molecule.

Equation 6.1 $Fe + H_2SO_4 \rightarrow Fe\ SO_4 + H_2$

As illuminating gas became popular after introduction by William Murdock in 1792 for lighting and cooking, it became apparent that a balloon could be filled with "lifting gas" from an illuminating gas pipeline. Illuminating gas was a mixture of hydrogen and carbon monoxide and would readily fill the balloon to provide good buoyancy but not as much as pure hydrogen. Equations 6.2 and 6.3 summarize Murdock's production of Hydrogen:

Equation 6.2   Coal + Steam $\rightarrow$ Carbon Monoxide + Hydrogen

Equation 6.3 $Carbon + H_2O \rightarrow CO + H_2$

The density of carbon monoxide is close to being neutral with air so the chemists of the day used slaked lime filters to take most of the carbon monoxide and carbon dioxide out of the illuminating gas. This provided better lifting force with nearly pure hydrogen. Equation 6.4, which is not balanced, shows the general approach for making relatively pure hydrogen by precipitating the oxides of carbon as calcium carbonate. Sodium and potassium solutions were also used to collect carbonates by caustic scrubbing.

Equation 6.4 $CaO + CO + CO_2 \rightarrow Ca\ CO_3$

Lime was known from the time man tamed fire and was widely used as mortar for building stone hearths and fireplaces. Heating limestone to red heat drives off carbon dioxide to produce calcium oxide or "quick lime." In England, the Norman invasion brought a new technology for building with stone, and lime burning became important to make strong mortar. Soon after William Murdock started to commercialize illuminating gas which also became known as "coal gas"or "town gas" it became common to use quick lime to absorb water vapor, carbon dioxide and carbon monoxide, to provide purified hydrogen. The lime industry had a new market.

PRESIDENT LINCOLN'S HYDROGEN AERONAUTS:

President Abraham Lincoln realized that horseback observations were often the best source of information regarding battlefront conditions. He had no doubt the side with the best intelligence would have a better chance at winning the Civil War. Both armies had brave men and enough gunpowder to win the war. He welcomed a way to make Union generals more aware of battlefield situations and complexities. Horseback elevation for surveillance, while better than walking or crawling, had serious limitations and dangerous drawbacks. An observer usually had to be within rifle or artillery range and the landscape usually made the field of view very small. This made it extremely difficult to know where and what the enemy strength might be or even where your own troops were.

6.1

In 1861, Lincoln appointed Thaddeus Sobieski Lowe as Chief of Army Aeronautics. In less than three years, Lowe and his civilian aeronauts logged more than 3,000 safe flights to report Confederate military movements and conditions. Lowe's *spy in the sky* program would evolve into the *space race* a century later. Hydrogen was the key to progress in both contests.

HYDROGEN GENERATORS:

Battle-front movements and conditions required rugged field generators to quickly produce hydrogen for the observation balloons that would carry an aeronaut with a telescope high enough to provide detailed information about Confederate troop and naval maneuvers. Lowe commissioned the emergency development of portable gas-tight canisters with numerous shelves to hold multitudes of iron filings that could be reacted with sulfuric acid. Hydrogen of adequate pressure could be produced at the rate that sulfuric acid was released on the iron particles. Lowe's portable hydrogen generators were designed to fit a standard Army wagon so they could be quickly delivered to new locations during advancements and retreats of the Union Army.

6.2

Lowe built five airships of various sizes and assigned one or more of twelve new hydrogen generators to each airship. About 1,200 square yards of silk were needed to build his largest airship, which was called the Intrepid. To fill the Intrepid's silk bag, about 32,000 cubic feet of hydrogen was released by reacting iron filings with sulfuric acid. Two hydrogen generators could fill the Intrepid in less than six hours.

The North's aeronauts would tether the observation balloons far enough from enemy positions to hopefully stay out of range of rifle and artillery fire. Lowe secured the best telescopes and taught his observers how to use trigonometry and geometry to determine distances so they could map the war fronts and Confederate options for battle maneuvers. He was determined to provide indispensable information from the advantage of high-altitude observation. Confederate troops and naval vessels would only have a few hours of activity at night before discovery by Lowe's aeronautical intelligence gathering.

Lowe had the reputation of being the most shot at man in the Civil War. He ordered all observation balloons to fly high and to be securely tethered to well-protected positions far out of rifle and artillery range. One of his aeronauts, John La Mountain, complained often about Lowe's requirement of being tethered so far away from the battle front and early in 1862 managed to make a free flight to track a group of Confederate troops. However, LaMountain traveled far enough to surprise a group of Union troops that were not informed of his mission. When he quickly descended to land, they thought La Mountain, who wore civilian clothes, was a Confederate spy and shot him and his unmarked hydrogen balloon. Although the hydrogen balloon quickly deflated it did not burn. La Mountain was the first recorded hydrogen aeronaut to die from gun fire. Friendly fire had caused the only casualty of a hydrogen aeronaut in a terrible war that ultimately killed more Americans than any other.

General Lee quickly realized the military significance of the observation balloons and launched makeshift hot-air balloons but lacked the funds to match the North's program. In order to respond to the North's hydrogen observation balloon advantage, Confederate leaders appealed for contributions of silk dresses. Soon seamstresses were finishing a great patchwork of silk dresses to make a hydrogen observation balloon.

The South had no portable hydrogen generators so the patchwork balloon would be filled with town gas in Richmond, Virginia and taken by a steamer down the James River or pulled by a rail locomotive down the York River Railroad to a place near where a General needed observation of enemy activities. One day it was being taken down the James River when the tide went out and stranded the steamer high and dry on a sandbar. Union troops quickly mounted an assault to capture the hapless crew and successfully took away the Confederacy's silk-dress balloon. General Longstreet said, "This capture was the meanest trick of the war and one that I have never yet forgiven."

LONG-RANGE FREE FLIGHT:

6.3 On March 14, 1899, Ferdinand Graf Zeppelin, a retired German military officer was awarded Patent Number: 621,195 for a Navigable Balloon. Zeppelin and an engineer named Theodore Kober formed a company that developed lighter-than-air concepts and built dirigibles called "Zeppelins" that had metal frames with a rigid skin within which very large lifting-gas bags were secured. Eventually giant diesel engines would power the Zeppelins and they could fly great distances and deliver heavy loads.

Zeppelin had been a military observer with the Union troops and had monitored the innovative progress of Professor Thaddeus Lowe's hydrogen generators and observation balloons. After retiring from military service in 1891, Zeppelin and Kober built a powered airship that used hydrogen as the lifting gas and first flew on July 2, 1900, three years before the Wright Brothers would fly the first heavier-than-air airplane at Kitty Hawk.

The Wright Stuff:

Orville and Wilbur Wright built their own gasoline engines to get enough power to enable a sustained flight with roll and yaw control. After their 57 second first flight at Kitty Hawk, South Carolina in 1903, they built two other powered aircraft before showing a more practical flyer with two upright seats in 1907. This convinced the United States Army Signal Corps to place an order for the first heavier-than-air military plane, which the Wrights delivered in 1909. By 1912 the Wright Model D had a 6-cylinder engine with a shorter wingspan and a top speed of 66 miles per hour. Their innovative pioneering had launched the industry that would build heavier-than-air commercial and military airplanes but they were disheartened because of legal and business problems. Wilbur Wright died of typhoid fever in 1912. Orville sold the company in 1916 and went back to inventing.

Count Zeppelin and Kober's flight tests continued and on August 5, 1908, while on an exhibition trip to raise funds, Count Zeppelin landed to fix some minor problems. After he left the tied down LZ 4 airship to get repair parts, it was demolished by a thunderstorm and burned. Zeppelin's company was ruined; it had no funds to continue. Surprisingly, the German public, sorrowed by the loss, contributed over 6,250,000 Reichsmarks as a grant to restart Count Zeppelin's work and he formed the Luftscjoffbau Zeppelin G.m.b.H. (company) on October 8, 1908. Count Zeppelin had won widespread support by the German public.

As World War I approached, the Zeppelin manufacturing plant became a war factory and made many dirigibles. At the beginning of WWI, Zeppelins could fly much higher and out of the range of airplanes. Zeppelins had become the *spy in the sky*. After turbochargers became available, fighter planes could fly high enough to successfully attack Zeppelins but only if they were near the fighter's air base and the attack did not take very long because their range was relatively small. Zeppelins could stay aloft indefinitely.

A new tactic was needed that would take advantage of the long-range flight advantages that Zeppelins offered. Zeppelins were loaded with additional fuel to extend their advantage of long-range travel. A German naval airship, L59, flew from Bulgaria with reinforcements for the war front in German East Africa. After flying about halfway the captain of L59 was fooled by a false order to turn back that was convincingly given by an English radio operator. The aborted trip covered about 6,750 km (4,194 miles) and showed that Zeppelins could be built to cross the Atlantic and fly back to Europe without stopping. U.S. Military strategists suddenly realized the potential threat by Zeppelins that could photograph or attack U.S. cities. Count Zeppelin died in 1917 and although he envisioned his Zeppelins as passenger and mail carriers considerable fear and hatred had developed because of their military uses.

Germany was forced by the Treaty of Versailles to make punitive war-reparation payments and restrict the size of airships that could be built. Steel cylinders or canisters that had been standard since about 1895 for holding hydrogen at 2,000 PSI were demanded as part of the war reparation payments. Many of these cylinders are still in service today and can be found in England, France, and the U.S.

John Alcock was a British bomber pilot who had been shot down in Turkey and imprisoned during WWI. After the war, he and Arthur Whitten Brown, who had also been shot down and imprisoned during WWI, joined forces with aircraft manufacturer Vickers to prepare a Vickers "Vimy" bomber to fly across the Atlantic ocean. They planned to fly over the 1900-mile course proposed in 1849 by Horatio Hubbell for the Atlantic Telegraph Cable.

Alcock and Brown were the first to fly across the Atlantic and collected the 10,000-pound prize that was offered for the daring accomplishment. Their1919 non-stop air travel across the Atlantic from Newfoundland to Clifden, Ireland was accomplished in 15 hours and 57 minutes, much of the time in the dark over cold, stormy, North Atlantic seas.

A Zeppelin could have flown the same course and returned without stopping. In 1924 a captured hydrogen Zeppelin, the LZ 126, was flown across the Atlantic for delivery to the U.S. Navy. The Navy changed the lifting gas to helium and renamed the airship ZR III Los Angeles.

In 1927, Charles A. Lindbergh completed the first solo nonstop transatlantic flight. Lindbergh flew his Ryan NYP "Spirit of St. Louis" 5,810 kilometers (3,610 miles) between Roosevelt Field on Long Island, New York, and Paris, France, in 33 hours, 30 minutes.

A passenger service hydrogen airship called the Graf Zeppelin was christened in 1928. Graf Zeppelin made 650 flights without any serious mishaps. More than 18,000 passengers were delivered safely during the nine years that the Graf Zeppelin flew. Compared to much slower travel across the seas by ships, or much shorter-range heavier-than-air passenger planes, passengers enjoyed 144 swift, nonstop, flights to and from Berlin across the Atlantic to Rio de Janeiro or New York.

Graf Zeppelin's travels amounted to more than one million miles or 40 times around the world including a 20-day cruise around the world in 1929 on a publicity flight and a trip to the North Pole in 1931. Graf Zeppelin was lifted by air displaced by sixteen giant "sausage" casings filled with hydrogen. One of the better German technologies that had been years in development was *sausage casings*. German craftsmen adapted this technology to make lightweight lifting bags for most of the Zeppelins that were built from 1900 to 1935. These bags were reinforced with cotton fabric and filled with hydrogen to atmospheric pressure. Over 800,000 ox-guts were required for the gas bag liners used in the Graf Zeppelin.

In 1940 the Graf Zeppelin, although still airworthy, was dismantled by order of Reich Marshal Goring. Goring's plans did not include peaceful passenger service to and from international cities. He planned to attack and subdue these cities.

REMEMBER THE HINDENBURG?

The 129th Zeppelin that was built was called the "Hindenburg." It was launched in 1936 and had safely crossed the Atlantic 21 times. It used Goodyear-manufactured gelatin-latex membrane to contain the hydrogen in the gas cells. Careful study should be made of the disastrous mishap that burned the Hindenburg in 1937. An eye-witness passenger reported events as follows on the fateful evening that the Hindenburg burned while attempting to dock at Lakehurst, New Jersey: "With my wife I was leaning out of a window on the promenade deck. Suddenly there occurred a remarkable stillness. The motors were silent, and it seemed as though the whole world was holding its breath. One heard no command, no call, no cry. The people we saw (on the ground) seemed suddenly stiffened. I could not account for this. Then I heard a light, dull detonation from above, no louder than the sound of a beer bottle being opened. I turned my gaze toward the bow and noticed a delicate rose glow, as though the sun were about to rise. I understood immediately that the airship was aflame ... For a moment I thought of getting bed linen to soften our leap (from 120 feet) but in the same instant, the airship crashed to the ground ... We leaped from the airship ... my wife called to me; ... took me by the hand; (and) led me away." (From the book "The Last Trip of the Hindenburg" by Leonard Adelt.)

This was an eyewitness report of the burning of an airship that carried a crew of 59. It had capacity for 50 passengers in individual cabins or for 70 passengers on day flights. On the evening it burned, the Hindenburg carried 97 persons.

Passengers had ornate individual cabins with shower baths, a clubroom for all with an aluminum grand piano, and a carefully insulated smoking room. The kitchen stocked two tons of the finest foods. Public rooms were large, decorated in the style of ocean liners of the day and they had windows that could be opened for fresh-air viewing of the grand scenes that unfolded as the giant airship sped along at the cruise speed of 78 mph. It was the largest airship ever built, with an 800-foot long aluminum frame filled with 7,200,000 cubic feet of hydrogen contained in 16 bags made of two layers of woven fabric with a gelatin-latex plastic film cemented between.

"The actual cause of the fire was the extreme easy flammability of the covering material brought about by discharges of an electrostatic nature ..."

OTTO BEYERSDORFF, HINDENBURG CREW MEMBER AND ELECTRICAL ENGINEER

After being launched in 1936, the Hindenburg had completed ten and one-half round trips between Germany and the United States before burning in 1937. Cruising across the Atlantic took 50 to 60 hours under constant power from four 1,200-H.P., V-16 Mercedes-Benz Diesel engines. Wooden propellers 20-feet in diameter were turned by these V-16 engines. The fully loaded range was about 10,000 miles or about 5 to 6 days at cruise speed.

Two 30-kilowatt Diesel-powered generators carried the electrical loads and a stand-by unit could deliver additional power if necessary. Passengers received the best food and drinks, the most modern conveniences, and the envy of other travelers because the Hindenburg sped past ocean liners, outran trains, and remained airborne for days or potentially weeks after other aircraft had to land and refuel. Telephones, electric lighting, and modern appliances served the crew and passengers.

Germany's Nazi Third Reich provided funding to build the Hindenburg. It was run by the Nazi Minister of Propaganda. Huge swastikas were painted on the tail fins and loudspeakers made Nazi propaganda announcements when the giant ship toured cities that it passed. Thousands of small Nazi flags were dropped to float down like tiny parachutes to thrill school children and others who watched the giant Zeppelin pass.

Before World War II, certain natural gas wells in the United States were the only significant sources of helium. Helium was extracted from natural gas produced from wells around Hugoton, Kansas. Although the Hindenburg was designed to use inert helium as the lifting gas, U.S. military authorities prevented exportation of helium to Germany. If the Hindenburg would have been filled with helium, would it have burned and crashed at Lakehurst, New Jersey?

Germany recognized the military possibilities of powered dirigibles and had used them in World War I. The Hindenburg type of airship represented considerable technical advancement and posed a much larger threat because it could fly to virtually any target, drop bombs or propaganda, and fly back to Germany without stopping.

After the Hindenburg burned, much speculation about sabotage entered the investigation. Was the disaster caused by lightening, sabotage, or something else? Nazi investigators were never convinced that the fire was caused by natural sources. The U.S. Government still holds strategic reserves of helium and closely monitors production and export programs, but the reasons for doing so have shifted from dirigibles and centered on the relative scarcity of helium and its myriad of applications ranging from use as an inert cover gas for welding to various heat-transfer applications.

Regardless of such speculation, translation of a letter handwritten in German on June 28, 1937, by Hindenburg crew member and electrical engineer Otto Beyersdorff states, "The actual cause of the fire was the extreme easy flammability of the covering material brought about by discharges of an electrostatic nature ..." NASA investigator Dr. Addison Bain has verified this finding by scientific experiments that duplicated the vigorous ignition by static discharge in the aluminum-powder filled covering material. Dr. Bain concluded that the Hindenburg would have burned even if helium had been used as the lifting gas.

On the fateful evening, reporters and camera crews gathered expecting to see a "high docking" in which the Hindenburg would be moored to a mast and secured with ground lines. Their cameras recorded what happened as the Hindenburg dropped lines to waiting landing crews and the events after the flames appeared. For a while the Hindenburg remained buoyant while flame spread along the outer skin. Obviously the hydrogen was still within the gas bags. Spectacular colors were produced from the burning skin of the giant airship. Hydrogen is about fourteen-times lighter than air. After the flames impinged the lifting bags, hydrogen rapidly vanished upwards and the airship started to gently fall in the first moments after the fire spread to additional hydrogen cells.

Flames from hydrogen combustion traveled 500-feet upward, far away from the crew and passengers in the cabins below. What fell to the ground with the passengers were burning shrouds from the exterior fabric, diesel fuel, and combustible materials that were in the cabins. Jumping or falling out of the descending Hindenburg caused most of the fatalities. Two were killed by burns. Flames continued to be supported by heavier-than-air fabric and diesel fuel that fell to the ground. The ground fire continued for hours.

Sixty-two persons from the Hindenburg lived through the disaster by being fortunate enough to ride the Hindenburg down and escape the flames and wreckage that fell to the ground. Most of these survivors were relatively unharmed.

"Remember the Hindenburg" should bring thoughts of the 200 persons in the landing-assist team who were below the Hindenburg reaching for mooring ropes when the Hindenburg caught fire. If the Hindenburg had carried the same amount of gasoline as the energy released by burning the 7,200,000 cubic feet of hydrogen ... the loss of life would have surely included many more passengers and the 200-member landing team would have faced a much larger "rain of fire."

Investigation of the Hindenburg disaster by the German engineers who analyzed the disaster in 1937, by the German Aerospace Research Establishment in the 1990's and by a team led by Dr. Bain of NASA proved that it was powdered aluminum in the flammable paint varnish that coated the infamous airship, not the hydrogen that initiated the fateful fire.

QUEST FOR LIQUID HYDROGEN:

Experiments by James Joule in 1845 and later refined by J.J. Thomson showed that a gas, as it slowly expanded from a high to a lower pressure, undergoes a change in temperature. Some gases are heated, but most are cooled by the expansion, depending upon the initial temperature. At room temperature, for most gases, expansion results in cooling. Compressed air starting from 273 K, (32°F) for example, will drop about 0.25K (0.45°F) for each atmosphere drop in pressure. Carbon dioxide will drop 1.5 K (2.7°F) for each atmosphere reduction in pressure. The inversion temperature is the temperature below which an expansion produces cooling. For most gases the inversion temperature is above room temperature, but for hydrogen, it is about 193 K (-80°F) compared to 273 K (32°F) as the water-ice point. In order to cool or liquefy hydrogen by gas expansion it must first be cooled below 193 K (-80°F) before it is expanded.

6.5

In 1895, a breakthrough occurred in the practice of gas liquefaction. Regenerative cooling was invented for the liquefaction process. Regenerative cooling utilizes a fluid as the coolant in a process in which the fluid is itself involved. For liquefaction of gases, it means that a gas that is cooled by the Joule-Thomson expansion process can later be used to cool the incoming compressed gas (to a temperature below the critical point if needed) before expansion.

Regenerative cooling was first introduced by Siemens in 1857 and was used by Kirk, Coleman, Solvay, Linde, and others in refrigeration apparatus. Within two weeks of each other in 1895, William Hampson in England and Carl von Linde in Germany obtained patents for equipment to liquefy air using Joule-Thomson expansion and regenerative cooling. Linde described his apparatus to physicists and chemists at Munich in 1895. A number of publications appeared that same year, among them one by James Dewar, a Scottish physicist and chemist who described his apparatus for liquefying air using regenerative cooling.

Hampson's process for liquefying air started by compressing air to 200 atmospheres. He then expanded it to one atmosphere, and passed the expanded, cooled air through a baffled heat exchanger to cool the incoming compressed air. Hampson had his new apparatus making liquid air at Brin's Oxygen Works by April 1896.

Linde's approach was more complex than Hampson's, but it was more efficient and more suitable for large-scale production of liquid air and separation of oxygen from nitrogen. Linde used two stages of gas compression, with precooling of the compressed air using a separate ammonia refrigeration system, and employed a coiled heat exchanger having three concentric tubes for regenerative cooling. The heat exchanger was insulated by a wood case filled with wool.

Regenerative cooling proved to be the technology needed to liquefy hydrogen. James Dewar used it to become the first to continuously liquefy hydrogen on May 10, 1898. Liquid nitrogen was first produced with which Dewar precooled gaseous hydrogen, then he expanded it through a valve into an insulated vessel, also cooled by liquid nitrogen. Expansion of the hydrogen that had been pre-cooled below the inversion temperature of 193K produced liquid hydrogen, from about 1 percent of the hydrogen used.

Dewar measured the density of liquid hydrogen at 0.07 kilogram per liter which is 1/14 the density of water and about 1/12 the density of kerosene or gasoline.

The insulated vessel Dewar used was a double-walled vacuum container flask that he developed earlier. These vessels became known as "Dewar flasks," or simply "Dewars." Storage and transportation of very cold liquefied gases such as oxygen, nitrogen, air, hydrogen, fluorine, and helium depend upon Dewar's discoveries of insulation by vacuum and reflective radiation barriers. Dewars are double-walled vessels with a vacuum in the annular space to minimize heat transfer by conduction and convection. The walls are silvered to reflect radiant heat. Dewar became very confident about storing and transporting liquid hydrogen in his vacuum insulated vessels, predicting that it could be handled as easily as liquid air. Cryogenic vessels based on Dewar's discoveries are used today to transport liquid hydrogen with very low loss rates.

Thus, before 1900, many of the major properties of gaseous and liquid hydrogen were known. Quantity production of liquid air, oxygen, and nitrogen stimulated many industrial developments. Suggestions for storing liquid hydrogen for releasing gaseous hydrogen to replace balloon loss and for propulsion engines closely followed Dewar's accomplishment.

Excitement about Dewar's accomplishment reached an obscure Russian schoolteacher, and about five years after Dewar's discovery, Konstantin Eduardovich Tsiolkovskiy (1857-1935) proposed to use liquid hydrogen and oxygen to fuel a space rocket.

Propulsion efficiency is sometimes called the Froude efficiency after William Froude (1810-1879) who first quantified it. He defined propulsion efficiency as the ratio of useful power output to the rate of energy input. Propulsive efficiency ($\mathcal{E}$) in flight becomes:

$\mathcal{E}$ = Thrust (Flight Speed)/Kinetic Increase of Fluid)

Propellers provide the most efficient propulsion at low speeds. Jet engines can maintain efficiency at higher flight speeds by producing higher exit velocities of expelled gases. By producing much higher exhaust velocities, rockets can maintain good efficiency at very high flight speeds.

Rockets carry the fuel and oxidant or a "monofuel" that is to be used for propulsion and do not have to extract air from the atmosphere for combustion. This provides the advantage of not requiring acceleration of the atmosphere to produce an equal and opposite thrust. Rockets can provide thrust in the vacuum of space or theoretically in the oceans.

Tsiolkovskiy was an avid reader and probably knew of earlier work in hydrogen balloons and William Froud's equation. Although he tried to make hydrogen balloons and metal dirigibles, Tsiolkovskiy's rocket contributions are theoretical as he did not attempt experiments to demonstrate his theories. He did not possess advanced academic credentials, yet he was recognized and accepted by eminent scientists for his contributions.

In 1891, Tsiolkovskiy sent a paper on the theory of gases to the Petersburg Physico Chemical Society where it was well received by the members, including Dmitri Mendeleyev, the famed Russian chemist. His most significant contribution is his theory of rocket flight propulsion that he developed from Newton's laws of motion. In its simplest form it formalizes what travelers had known for ages — travel light, expend a lot of fuel energy, and you can go faster. Tsiolkoskiy theorized that the velocity of a rocket would be expressed as:

Equation 5.5 $$V = V_j \; \mathcal{L}_n \frac{Mo}{Me}$$

Where V is the maximum velocity of the rocket in gravity-free, drag-free flight, V_j is the rocket exhaust jet velocity. The natural logarithm $\mathcal{L}_n$ of the ratio of Mo (the initial or full rocket mass), and Me (the empty rocket mass).

Tsiolkovskiy's equation shows that a rocket's achievable velocity, V_{max} is directly proportional to the rocket's exhaust velocity. The exhaust velocity is essentially constant for a given rocket design, propellants, and operating conditions. Exhaust velocity depends upon the amount of heat energy released during combustion, the combustion pressure, the combustion products, and the nozzle for directionally expanding the gases.

The second term of the Tsiolkovskiy equation, $\mathcal{L}_n$ (Mo/ Me), involves two masses differing only in the amount of propellant (fuel and oxidizer) expended. The initial or full mass includes vehicle structure, tanks, engines, controls, guidance, propellants and the payload. The empty mass is the initial mass less the mass of propellants that have been expended. During operation, continuous burning of propellant and expulsion of exhaust causes the total mass of the vehicle to continuously decrease, starting with Mo and ending with Me. Tsiolkovskiy derived his equation of flight based on the conservation of momentum, integrating, and using the initial and final conditions of the rocket as bounds to obtain his equation.

It shows why rocket designers try to make the vehicle's structure, engines, guidance, and controls as light as possible because this and the payload become the empty mass. Liquid hydrogen and oxygen stored in dewers were the lightest readily produced propellant and would provide high-energy release to produce high exhaust velocity.

Tsiolkovskiy, in 1903 wrote his findings in terms of the velocity of the exhaust emerging from the nozzle. Similarly, Professor Goddard and others in the United States, expressed rocket performance in terms of the measured quantities: thrust and mass flow of the propellants. The thrust divided by the mass flow of propellant was defined as the specific impulse.

HYDROGEN ROCKETS:

6.6 Typical exhaust velocities for liquid propellant rockets range from 2000 to 4500 m/s (6500 to 14,700 ft/sec.). The V-2 had an exhaust velocity of about 2200 m/s (7200 ft/sec). High energy propellants give exhaust velocities in the range of 3000 to 4500 m/s, and the liquid hydrogen-oxygen combination is in the upper part of this range.

World War II brought competing secret programs to quickly produce rockets, radar, jet engines, atomic bombs and many other inventions that are less known. American rocket pioneer Robert Goddard tried to get the U.S. Military interested in rockets that could go to the moon or anywhere on Earth but his suggestions were rejected. However, captured German rocket scientists referred to Goddard's 200 patents as "the blue prints" for what would be 1,100 German's V-2's that demolished British Manufacturing centers, and terrorized civilians.

At the end of WWII, the Russian Army captured a group of German scientists who planned to make a cryogenic hydrogen powered space plane that would fly at high speed across the Atlantic to bomb American cities. Information from the super-secretive Soviet Union stopped. In order to know more about this and other threats that the Soviets posed in the 1950's, President Dwight D. Eisenhower ordered U.S. reconnaissance planes and camera balloons to fly over Russia.

In the Cold War that escalated, high-flying U-2 spy planes took pictures of new hydrogen liquefaction plants being built in the Soviet Union. U.S. Military analysts suspected that the captured German scientists under Soviet control were trying to complete their plans to build a cryogenic hydrogen fueled airplane to attack the U.S. Another emergency technology development program was given to Clarence L. "Kelly" Johnson at the sprawling Lockheed Aircraft complex in California. Kelly's Skunk Works had produced the famed U-2 that took the alarming pictures. Now he was called on to beat the Soviets with a cryogenic hydrogen powered high flyer. In May 1960, Francis Gary Powers was shot down and the wreckage of his U-2 was captured. Soviet rockets had improved rapidly to provide greater range and accuracy for hitting a high altitude target.

Kelly planned to build a hydrogen flying-wing that would skim the atmosphere at 100,000 feet and Mach 2 velocity (1,584 MPH). The Air Force secretly awarded $96 million to Lockheed for "Project Suntan" to build the prototype plane and a liquid hydrogen production plant. Soon the Skunk Works became the Free World's largest producer of liquid hydrogen for airplane fuel, creating 200 gallons a day, essentially with Dewar's 1898 technology.

Kelly's Suntan Hydrogen Aircraft was designated as a delta-wing vacuum bottle as big as a pair of B-52s. But when the Soviets launched Sputnik and it became apparent that Suntan would be no match for rockets that could orbit the globe at 17,000 MPH to quickly spy or deploy weapons on any target. The Space Race had started. High flying, air-breathing aircraft did not go high enough or fast enough.

The time for the long-range intercontinental ballistic missile (ICBM) race had arrived because a dreaded nuclear war head could be delivered anywhere on Earth in about one hour. Advanced military aircraft that resulted from an estimated $200 billion development cost that had been spent after WWII could not match the threat of unmanned nuclear tipped ICBMs that could deliver thousands of times more lethal nuclear destruction than the bombs dropped over Japan to end the war.

The U.S. and U.S.S.R. would spend as much as each could, as fast as possible, to build vast armaments of nuclear ICBMs and supporting armies and navies. The nuclear armament race produced a stand-off, and a bankrupt Soviet Union. Both superpowers had rapidly expended their once vast oil reserves. U.S. oil reserves had plunged to less than 3% of world reserves.

MANNED SPACE RACE:

6.7 The Soviet artificial satellite named "Sputnik" was launched on October 4, 1957. "Sputnik" orbited the Earth every 98 minutes. It weighed about 183 pounds and abruptly changed the scientific, military, and political atmosphere. *Spy in the sky* had entered gravity-free space at 17,000 mph. The U.S.S.R. scientists were ahead in the space race.

U.S. Military experts had counted on getting into the air to see what the enemy was doing and believed they had the upper hand since the time of the U.S. Civil War. But Sputnik cast serious doubt about who would have the upper hand in future surveillance. After Sputnik beeped for about a week and burned on returning to Earth's atmosphere, the Soviets launched a dog into near-earth orbit.

President Dwight D. Eisenhower quickly ordered development of spy satellites to improve on intelligence that had been developed by U-2 flights over the Soviet Union. The Corona spy-satellite program produced over 800,000 satellite images between 1960 and 1972. The first spy-satellite flight over the Soviet Union produced more images than all of the U-2 flights.

President John F. Kennedy continued to press for the space advantage and delivered a speech on May 25, 1961 that rallied the U.S. to land a man on the moon within a decade and safely return to Earth.

"...if we are to win the battle that is now going on around the world between freedom and tyranny, the dramatic achievements in space which occurred in recent weeks should have made clear to all of us, as did Sputnik in 1957, the impact of this adventure on the minds of men everywhere... Now it is time to take longer strides-time for this nation to take a clearly leading role in space achievement, which in many ways may hold the key to our future on Earth. ...we have never made the national decisions or marshaled the national resources required for such leadership. We have never specified long-range goals on an urgent time schedule... Space is open to us now; and our eagerness to share this meaning is not governed by the efforts of others. We go into space because whatever mankind must undertake, free men must fully share... I believe that this nation should commit itself to achieving the goal, before this decade is out, of landing a man on the Moon and returning him safely to the Earth. No single space project...will be more exciting, or more impressive to mankind, or more important...and none will be so difficult or expensive to accomplish...".

The U.S. mission to the Moon would be propelled by the Saturn V hydrogen rocket developed under the direction of Werner von Braun, the German rocket scientist who was previously credited for applying Dr. Goddard's technologies to accomplish over 1100 V-2 rocket attacks on England. Saturn V was a three-stage rocket over 36 stories tall. Saturn V was propelled by liquid hydrogen in the upper stages. Boeing was selected to build the first stage. North American Aviation contracted to build the second stage. Douglas Aircraft would build the third stage. Rocketdyne Division of North American Aviation would build the engines for all three stages. Five F-1 engines powered the first stage, five J-2s the second stage and one J-2 the third stage. The dependable Saturn V would be called on for 32 launches without failure.

The first manned Saturn V delivered three Apollo 8 astronauts into Moon orbit in December of 1968. On 20 July 1969, Apollo 11's lunar landing module delivered Commander Neil Armstrong and Edwin Aldrin to the "Sea of Tranquility" landing site on the Moon while Michael Collins orbited above in the command vehicle. As he first stepped on the Moon, Neil Armstrong said, "That's one small step for man… one giant leap for mankind."

The Soviets had a series of Moon rocket failures during the test phase but attempted to gather the first samples from the Moon and sent a robot rocket named Luna 15 two days before Apollo 11 lifted off. It arrived shortly after U.S. lunar astronauts Neil Armstrong and Buzz Aldrin took their first walk on the Moon. Luna 15 crash-landed. It could not return to Earth and remains on the Moon as the first extraterrestrial junk yard.

Five more Apollo missions landed astronauts on the Moon. Apollo 17 was the last mission to the Moon. Eugene Cernan commanded the flight, Ronald Evans was the command module pilot and Harrison Schmidt was the lunar module pilot. Apollo 17 landed on 11 December 1972 and returned to the Earth on 19 December 1972. The six missions safely brought back about 400 kilograms of Moon samples and volumes of scientific data and photographic records of the Moon.

In January of 1972, President Nixon announced a new NASA program that he described as a Space Transportation System (STS) and eventually it became more commonly known as the Space Shuttle Program. U.S. Space Shuttles were designed for repeated travel to and from space. The shuttle and solid fuel booster were designed to be reused far into the future.

Columbia would be the first U.S. Space Shuttle. Columbia with John Young commanding and Robert Crippen as pilot launched into space on 12 April 1981. With a launch weight of 219,258 pounds the Columbia would make 37 orbits around the Earth and return two days and 6 hours later after traveling over one million miles. Columbia proved the viability of the concept of launching and recovering the solid propellant rocket boosters and the main spacecraft for reuse. Space travel looked safe. But "looks" were deceiving.

O-RING LEAK CAUSES CHALLENGER DISASTER:

Challenger, a space shuttle that followed the successes of the Columbia was scheduled to launch early in the morning of 27 January 1986. It was 36 degrees Fahrenheit, about 15 degrees colder than the coldest previous launch. At 0.678 seconds after liftoff smoke was spurting from the right solid rocket booster. O-ring seals were eroding and at 58.788 seconds a steady flame impinged

6.8

on the super cold liquid fuel and oxidant tanks. Flames at 5,800°F from the leak in the Solid Fuel Rocket Booster expanded as the O-ring seal burned away. The growing blowtorch from the failing booster blasted into the enormous hydrogen and oxygen tanks and quickly spread to weaken the structural supports for these tanks. About 73 seconds after launch, after gaining an altitude of 46,000 feet, and Mach 1.92 velocity, a huge fireball engulfed the Challenger as the breached hydrogen and oxygen tanks dumped enormous energy into the fatal inferno.

Engineers of the Solid Fuel Rocket built by Thiokol had previously discovered that ambient conditions below 53 degrees Fahrenheit would cause the O-rings to shrink and be too stiff to seal adequately. Managers at NASA and Thiokol ignored or overruled the warning and decided to launch the Challenger into the cold morning air at 36 degrees Fahrenheit. America's greatest space-launch disaster could have been prevented, but the warning report was ignored and the decision to re-engineer the seals or to wait until warmer ambient conditions necessary to save the lives of the seven Challenger astronauts was disregarded.

The Challenger disaster caused important changes. The infamous o-rings were re-engineered and replaced. NASA adopted a pre-launch system that actively solicits information from the astronaut crew, engineers, and operations groups. Ultimately the final launch decision is made by one of the astronauts on the flight crew and not by publicity-minded politicians, cost accountants, or managers at NASA, Thiokol, or any of the other contractors that supplied components.

The Break up of Columbia upon re-entry into Earth's atmosphere, 17 years after the Challenger accident, on February 3, 2003, reminds us that there are many hazards associated with space flight.

SPIES IN THE SKY:

Currently spy satellites "hear" more and "see" far better from space than Lincoln's Civil War Aeronauts could with their best telescopes. A security analyst with the Federation of American Scientists, Tim Brown commented, "Our best approximation is 10-centimeter (3.9 inch) resolution." Not only is the resolution amazing but the most modern spy satellites can produce images at night or when there is cloud cover. Compared to the occasional spy plane flyover, a space satellite can be arranged to pass over every few hours or less to get a progress report.

6.9

What the satellites report is that virtually all of the human population now depends upon expending ever increasing amounts of fossil resources. The rate of fossil fuel burning has more than tripled since the time that President Eisenhower ordered spy flights over the U.S.S.R. and China. And the global climate is changing.

At the beginning of the Oil Age, the U.S. had about as much oil as Saudi Arabia but today U.S. reserves are down to about 2.2% of global reserves. U.S. demand for oil is about 27% of global production and the U.S. GNP is about 25% of the global GNP. From the beginning of the Oil Age until being challenged by the Soviet Union during the "Cold War", the U.S. was the world's largest oil producer.

NOW THAT WE CAN SEE SO WELL, WHAT DO WE SEE?
"This thing we call (civilization) —all these physical and moral comforts, all these conveniences, all these shelters,... constitute a repertory or system of securities which man made for himself like a raft in the initial shipwreck which living always is—all these securities are insecure securities which in the twinkling of an eye, at the least carelessness, escape from man's hands and vanish like phantoms."
- Ortega y Gasset

Analysis of the information gained about "who, when, and where," and application of what has been known through the centuries about human nature as to "why" has shown that humans can be expected to allow the advantages that facilitated the Industrial Revolution to be wasted. The U.S.S.R. used central planning to find ways to extract oil faster than the U.S. The U.S.S.R. became the world's largest oil producer. Much of the oil reserves of the U.S. and U.S.S.R. were expended, however, in an arms race that far overbuilt conventional armies, nuclear warheads, nuclear submarines, and other weapons of mass destruction. China and India with much larger populations are vying to use as much of the global oil reserves as did U.S. and U.S.S.R to join the temporary oil boom and arms race.

We must realize that the buyer's market for fossil fuels that facilitated the Industrial Revolution has changed with Peak Oil to a seller's market. The ability to meet increasing demand for oil with increased production is gone.

Why not use the great technologies that the Industrial Revolution has provided to change to renewable fuels? This must be done very soon and carefully to keep from defeating the progress that has been made by fossil fuel expenditures in agriculture, architecture, transportation, and commerce.

LAUNCHING THE RENEWABLE RESOURCES REVOLUTION:

Successful revolution is a state of mind that embraces facilitating inventions to overcome habits, superstitions, hunger, tyranny, and so many other human impediments.

Crisis looms for the world economy. It is a crisis produced by 6.5 billion people dependent upon burning over one million years' of carbon-based fossil accumulations each year. We are dependent upon diminishing supplies of fossil resources. We could try to make an economy as Civilization did prior to exploitation of fossil resources by again becoming dependent upon vegetation and animal resources. But the growing human population is over six times larger now and we have greatly increased expectations for far more energy-intensive dwellings, consumer goods, and travel.

CHAPTER 7.0

Green plants, despite all the great things they do, cannot make enough biomass to provide all of the fuel, food, and fiber that we now demand. Green plants fed to animals would supply even less fuel, food, and fiber. Adequate fuel for goods production, transportation, and modern lifestyles poses a difficult supply problem. We now use the fossil equivalent of about 200 million barrels of oil each day.

We could harvest all of the biomass that could be sustainably produced without fossil expenditures and convert it into a fuel such as methanol or ethanol. But it would not equal the amount of fossil fuels now required. And doing so would sacrifice food, fiber and wood-based building materials.

First priority:

Biomass that rots or burns, however, represents an important opportunity. Carbon dioxide and methane are greenhouse gases that must no longer be released into the atmosphere as waste gases. Biomass of virtually every kind can be converted into carbon products, hydrogen, and soil nutrients.

Because methane is about 27 times more harmful as a greenhouse gas than carbon dioxide it should be used as a preferred source of carbon and hydrogen. When this is not practical, methane should be burned in an engine or utilized in a fuel cell instead of being released from anaerobic decay of biomass. Utilizing renewable methane as a fuel instead of fossil coal, oil, or natural gas can substantially reduce greenhouse gas pollution.

TESI CRISIS RESPONSE SYSTEM:
The Hydrogen Internal Combustion Engine (ICE) powered Total Energy System Innovation (TESI) provides immediate support for rescuing disaster-stricken communities, long-term production of renewable resources, and exemplary methodology for doubling energy-utilization efficiency.

One way to initiate the state of mind needed for successfully launching the Renewable Resources Revolution is offered by the Crisis Response System of Figure 7.1. The simplest Crisis Response System includes two TESI trailers of equipment to help overcome a disaster caused by an earthquake, flood, famine, fire, or tornado. Mounted on one trailer is a TESI, consisting of a hydrogen engine-generator, cooking center, and water condenser. A second trailer brings a Waste Converter that produces renewable methane, hydrogen, carbon, and soil nutrients from sewage, garbage, waste wood, disaster debris, and other organic materials.

7.1

FIGURE 7.1

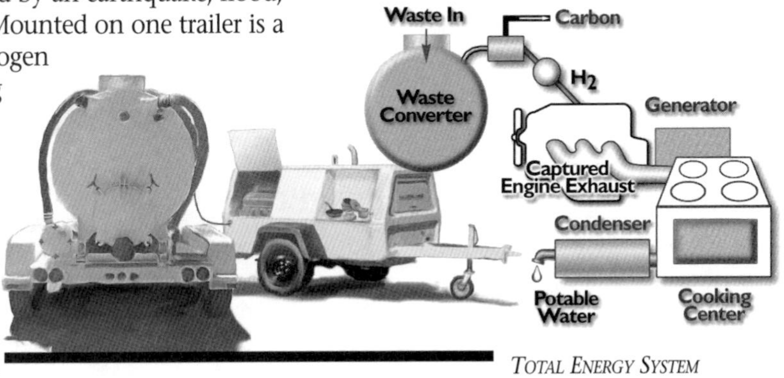

TOTAL ENERGY SYSTEM INNOVATION (TESI) Combining the waste Converter and Hydrogen ICE in a TESI provides rapid response to disasters including the energy crisis.

Hydrogen produced by the Waste Converter is burned in the ICE, which drives the electricity generator. *The community has electricity.* Steam exhausted by the engine is circuited through heat-exchanger passages in the oven and/or cook top to prepare food for the disaster relief teams. *Disaster relief requires hot meals to re-invigorate tired workers.* After supplying heat for cooking food, the steam is routed through a heat exchanger to provide *warm air for washing and drying clothes and warming emergency shelters* such as tents. Removal of heat from the steam causes it to condense into nine pounds of distilled water per pound of hydrogen that enters the engine. *The community has electric lights, refrigeration for storage of medicines and food, hot meals, warm shelter, clean clothes and distilled water.*

After the relief operations, the rescue TESI can be transformed into a Total Energy System for sustainable economic development. *This enables the community to start achieving energy independence.*

Sewage, garbage, and organic wastes produced by the disaster are loaded into the Waste Converter to produce fuel for the Hydrogen ICE. After the disaster relief phase of operation, the community can expand production of renewable hydrogen for energy-conversion applications by disposing of sewage, garbage and crop wastes in the Waste Converter. *First the community overcomes the disaster and then overcomes the economic crisis of dependence upon shrinking supplies of polluting fossil fuels.*

RENEWABLE CARBON FACILITATES WEALTH EXPANSION:

7.2 Carbon that is sequestered from the Waste Converter can be used to produce renewable materials and durable goods that are superior to conventional products. Carbon fibers can be produced to reinforce products that are stronger than steel and lighter than aluminum. Carbon fiber components will make future inventions far more reliable and economical than conventional products.

The Carbon Age we are entering will be defined by materials with better chemical and physical attributes for nearly every application now served by iron, steel, aluminum, magnesium, nickel, cobalt, and silicon. Carbon products that provide these improvements will be manufactured with less energy content than the materials that have launched the Industrial Revolution. The Solar Hydrogen Civilization will embrace lighter, stronger, materials that are manufactured with less environmental impact. Sequestering carbon from biomass wastes and hydrocarbon fuels will overcome greenhouse gas pollution from carbon dioxide, methane, and organic gases by extracting carbon before these gases are generated and released by decay or combustion. Carbon nanomaterials and devices will provide greatly improved microelectronics, medical technologies, and instrumentation.

Tanks for storing hydrogen at energy densities equal to gasoline will be made from carbon. Such tanks are far safer than conventional gasoline tanks. Carbon can be made into crucibles and heat exchangers that can be used to melt and cast the super alloys used in jet airplane engines. Carbon can be made into diamond plating that provides greatly improved wear and corrosion resistance of tooling for manufacturing other components. Carbon will be used to make fuel cells and electrolyzers. Activated carbon filter media will purify water, air, and industrial process fluids.

Although in most areas the amount of hydrogen gained from Waste Converters will not be enough to replace all of the energy now released by burning fossil fuels, the sequestered carbon will be used to produce devices for conversion of sufficient renewable energy to provide sustained prosperity. Sequestered carbon will be used to build the solar dish gensets, wind machines, wave generators and energy efficient transportation equipment to make the community economically sustainable and prosperous.

Communities that rebuild after disasters with renewable energy and materials will have confidence to prosper without pollution. Transition to renewable resources will allow these communities to escape economic inflation, energy shortages, and hardships caused by dependence upon fossil energy.

RED CROSS, U.N., SALVATION ARMY, AND CHURCHES CAN SAVE MONEY AND OVERCOME MORE DISASTERS:

By bringing Waste Converters and Hydrogen ICE powered **7.3**
TESI(s) to facilitate disaster relief efforts, Red Cross, the Salvation
Army, and similar humanitarian organizations can more usefully
redirect much of the cost required to buy diesel fuel for conventional
electricity generators. The community will have positive incentives to collect debris and wastes produced
by the disaster. Waterborne diseases and pollution will
be curbed as would-be pathogens are converted into
hydrogen, carbon, and soil nutrients that restore productivity of crop-lands.

Restoration of trace elements on crop-lands will
reverse the adverse practice of imprisoning trace elements as garbage and sewage sludge that are buried in
landfills. By today's wasteful practices in industrialized
countries, soil that grows crops eventually becomes
depleted as essential trace minerals become buried in
landfills that are designed to imprison the garbage and
sewage sludge for centuries. In less developed areas,
wastes including ashes are washed down streams and
rivers to the ocean. In both cases the crop-lands are depleted.

Improving the soil productivity and the nutrient value of crops
that are grown after restoration of trace minerals is an important economic driver of the Renewable Resources Revolution. Persons who
embrace this peaceful, healthful, wealth-expanding revolution will eat
better, have more energy, and live longer.

One of the illogical representations that tries to dismiss
Civilization's dependence upon burning Earth's fossil deposits and the
resulting environmental degradation claims that it is natural
for geological events like volcanoes or forest fires to release
more carbon dioxide than the amount sourced by human
activities. Regardless of the comparative amounts of carbon
dioxide release, it is imperative that we quickly do something
about what we can control. It is particularly imperative in reference to atmospheric records of increasing methane and carbon dioxide concentrations.

Increasing concentrations of atmospheric carbon dioxide
and methane signal the inefficiency of our economy and the
opportunity cost of sacrificing carbon that could have been
made into extremely valuable durable goods.

During at least the last 160 thousand years, all the volcanoes, forest fires, erosion of continental soils, prairie fires, and
rot of oceanic biomass combined, established carbon dioxide
levels that varied little from an average of about 280 ppm. This
has been rapidly and dramatically exceeded by carbon dioxide levels produced during the last 100 years. Carbon dioxide levels follow seasonal variation but have ratcheted upwards to 330
ppm. Figure 6.2 illustrates the correlation of the consumption
of fossil fuels and the uptake of carbon dioxide during the

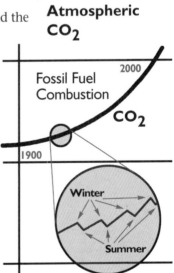

FIGURE 6.2

Atmospheric CO_2

Fossil Fuel Combustion

2000

CO_2

1900

Winter

Summer

growth seasons of photosynthesizing plant life to produce the ratcheting effect. As fossil fuel consumption increased, global carbon dioxide levels increased. Green plants temporarily utilize a portion of the atmospheric carbon dioxide to produce plant tissue but after the growing season much of the carbon dioxide reappears as plant tissues rot or burn, and the carbon returns to the atmosphere as carbon dioxide or methane.

THE LAW OF THE MINIMUM:

7.4 Renowned physician, the late Dr Robert Zweig often noted that Civilization is accountable to the "Law of the Minimum" as is everything else in the natural world. In any complex interactive system, an essential ingredient that falls below the minimum requirement will cause the whole system to collapse. If farmers deplete the aquifers in irrigated desert areas or cannot afford the energy to pump water, then all the fertilizer, know-how and market demand is immaterial. Similarly if cities run out of healthful air, all the investment in infrastructure, housing, roadways, factories, bureaucracy, education, etc. is at risk if not wasted.

Hydrogen is the unique fuel that can offer fundamental distributed energy sufficiency while satisfying the healthful air requirements of the Law of the Minimum. President Theodore Roosevelt promulgated representative government oversight on anti-trust, food and drug purity, and wilderness preservation. A century later these policies are still part of the core values of this country and our perception of what is right and necessary to preserve essential minimums for humans and other life forms.

Enduring policies are seldom built on abstractions, logic, or analysis but on what is perceived to be real. The motoring public perceives gasoline to be the cheapest fuel.

But in reality U.S. taxpayers annually provide $26 billion in subsidies for fossil fuels and an additional $19 billion for nuclear power. More than $54 billion subsidizes military protection of the Persian Gulf oil, which means that we may have been paying the equivalent of $70 per barrel for the last 20 years even before additional billions were spent for the Gulf War and Terrorist Wars. Indeed, Congressman Bill Alexander estimated that when the Desert Storm military costs are included in the cost of gasoline, the true cost of gasoline was elevated another $5.50 per gallon or more than $200 per barrel (see appendix-page 231).

War that continues in Iraq has more than doubled the true cost of gasoline for American taxpayers to more than $11.00 per gallon.

Every ten years since 1900, the amount of petroleum burned by the world's transportation sector has doubled! We cannot afford to continue burning the opportunity to produce present and future inventions of durable goods that can be made from the world's carbonaceous fossil reserves.

Health costs related to caring for diseases caused by, or aggravated by fossil fuel production and combustion continue to increase. Recent epidemiological studies reveal that citizens living in heavily polluted areas show increasing signs and incidence of chronic lung disease and certain cancers. Health-care dollars spent to care for these patients could serve a greater societal need by preventing the diseases. Solar hydrogen technology will go far in preventing diseases due to airborne contaminants while facilitating adoption of a sustainable economy.

OVERCOMING THE LOOMING WATER CRISIS:

In order to feed and clothe Earth's enormous human population, **7.5** farming has evolved into an amazingly productive enterprise that converts fossil fuels and energy-intensive fertilizers into foods and fibers. Fossil fuels are used to power the plants that liquefy air to collect nitrogen that is forced to react with hydrogen and form ammonia. Phosphate rock is reacted with an acid to provide a soluble phosphate. Potassium is prepared in energy intensive recipes for quick growth of irrigated crops. Fossil-fueled engines power pumps to deliver irrigation water from depleting supplies of glacial-age water reservoirs.

About 95% of the freshwater in the U.S. is in underground aquifers that were formed by erosion and glacial action that plowed alluvial sand, rock, and gravel into lower areas of the continent. These beds of rock debris provided excellent underground lakes of water-filled sand and gravel deposits.

Almost all of the aquifers that supply water for intensive farming operations are being rapidly depleted. Illustrative of the situation and growing problem is the Ogallala aquifer that extends from South Dakota to Nebraska, across Kansas, and the Oklahoma panhandle, to western Texas. Intensive farming operations that depend upon irrigation have depleted this aquifer at the rate equivalent to the annual flow of 18 Colorado Rivers. Ogallala aquifer depletion continues at the rate of 12 billion cubic meters per year (3.17 trillion gallons) more than rainwater replenishes it. In addition to this depletion, it takes enormous amounts of fossil energy, to pump this equivalent of 18 Colorado Rivers to irrigate the fields above this aquifer.

Similar problems exist around the world. China has rapidly increased farm yields by energy intensive application of commercial fertilizers and irrigation practices that are rapidly depleting local aquifers. India is doing as much as possible to increase farm productivity by energy intensive farming and irrigation. South American farmers are determined to supply the world market with meat, grain, cotton, fruits and vegetables that require irrigation and fertilizers.

In order to produce the foods and fibers that the growing market demands, much better utilization of fresh water is required. Growing plants in hydroponic conditions provides opportunities for reduced water and fertilizer consumption. Drip irrigation including directed surface drip and sub-surface drip offer considerable water savings per ton of plant production. Subsurface drip distribution provides numerous advantages including much less salting due to water deposition of dissolved solids, lower energy costs, and the opportunity to reduce or eliminate weeds by not watering them.

Farmers who now depend upon marketing crops that require more water than the local rainfall supplies have opportunities to utilize their lands to convert solar, wind, and biomass waste resources into renewable hydrogen and/or electricity for supplying cities with energy and renewable carbon. Crops must be increasingly selected on the basis of sustainable farming practices.

Considerable pollution and waste of ground water has also resulted from oil production practices. Most of the oil now available for 21st century markets was produced from the foodchain derivatives and remains of organisms such as microscopic marine plants called phytoplankton that were buried along with inorganic sediments on ancient sea floors some 60 to 600 million years ago.

Ordinarily, phytoplankton living in seawater would be eaten by other organisms ranging in size from euphausiids (small shrimplike "krill") to blue whales, the world's largest animals. A large blue whale might be 110 feet long, weigh 150 tons, and eat 4 tons of krill and phytoplankton each day. At times in geological history, however, the rates of phytoplankton growth greatly exceeded the rate of consumption by other organisms and organic residues of phytoplankton and various species of the food chain that depend upon the phytoplankton accumulated in sufficient quantities on the ocean floor to use up the dissolved oxygen. This produced anaerobic conditions in the decaying ooze and seawater near the bottom of the ocean.

Organic materials generally provide excellent thermal insulation as does silt, trapped gases and water produced by partial decay of organic materials. If these anaerobic residues were then buried sufficiently by more sediment, the temperature would increase in the decaying ooze because more heat normally passing from the Earth's hot core to the surface would stay within the insulated strata. Sufficient temperatures developed within such insulated strata to kill bacteria and cook the remaining organic mater even though much cooler seawater existed over the sedimentary strata of organic residues. In some settings these conditions existed in underwater canyons, valleys and other geological topographies that were conducive to delivery of the petroleum liquids and gases that formed into adjacent rock formations.

In order to produce a retainable oil and/or natural gas deposit, the petroleum being formed had to migrate into permeable porosity of a stable host rock formation. Host rock formations such as porous sandstone and limestone beneath impermeable layers served the purpose as oil displaced seawater within interconnecting pores. The impermeable layer over permeable sandstone or limestone trapped petroleum that rested over more dense seawater within lower zones of interconnecting porous networks. Through the eons of high temperature cooking, the seawater and petroleum produced much saltier and denser brines.

As the mountains of North America eroded and filled the sea with gravel, sand, and silt, the shoreline moved away from oil formations that were once at the bottom of the sea in Nebraska, Kansas, Oklahoma, Texas, and Louisiana. Oil found in these areas today is usually above considerable seawater that has been trapped in the lower pores of host rock formations for 60 million years or more.

Production of oil and natural gas is usually accompanied by various amounts of oil brine.

By 1916 Kansas had decided to protect fresh water and productive soils by enacting laws that required oil leaseholders to prevent oil and brine refuse from contaminating fresh water and soils. This was in response to the practice of operating "evaporation ponds" where anything that did not go to market was discarded. Evaporation ponds were dug in the soil over the local freshwater aquifers. They were filled with brine that was increasingly pumped along with petroleum values from the formations deep below. Brine ponds had to be fenced because brine kills animals that drink it and kills plants that it contacts.

Contrary to their name, the evaporation ponds did not evaporate the poisonous salts. What evaporated was water but not the offensive salt brines that killed animals and crops. Gradually poisonous salts that were discarded in surface ponds would diffuse downward along with rainwater into the fresh water aquifer. Poisonous brine seepage into fresh water supplies would eventually contaminate millions of acre-feet of glacial age water.

In 1966 the author bought a rural school complex that had been closed in favor of busing the school children to a larger school, supposedly for a less expensive education. The school complex consisted of beautiful brick buildings including classrooms, gymnasium, cafeteria, and administrative offices. I converted this complex into engineering offices and a manufacturing plant to produce equipment for making hydrogen and carbon from farm wastes, drip irrigation systems, and total energy systems for efficiently producing the electricity and heat at cascading temperatures for various applications ranging from cooking, curing concrete products, crop drying, water and space heating.

Subsurface drip irrigation reduces water requirements 60% to 85% by reduced evaporation, reduced watering of weeds, and reduced plant distress due to salt accumulations in the soil.

Suddenly in 1974 the well water that had been excellent since the well was dug was invaded by poisonous oil brine. Piping throughout the complex was poisoned and corroded. Water-cooled extrusion equipment was ruined.

An irrigation well that supplied water for my drip irrigation demonstrations was poisoned next. I contacted the nearby oil lease operators and they claimed that my problem was with much larger previous operators who had left oil brine in large evaporation ponds. After studying the problem I demonstrated how fresh water wells could be operated to create overlapping cones of depression to prevent further advances of the poisonous brines through the glacial-age public aquifer. But this required a lot of expense to prepare the

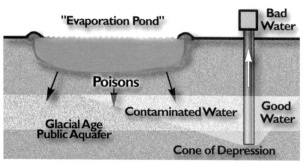

Hydrogen for the farm home, pickup, cars, and equipment engines along with oxygen for the oxygen/hydrogen replacement for the store bought oxygen/acetylene cutting and welding torches.

Wind & Solar

Hydrogen Generator

Hydrogen Powered Vehicles & Equipment

...ells for operation with very ...corrosive brine contaminants. And, I had no place to dispose of the poisonous water that had to be discarded without contamination of surface soils or the freshwater aquifer below.

...ers at one of the larger oil ...ies first indicated that they ...uld provide a disposal well so ...the offending brines pumped from the cones of depression could be returned to their original oil formation ...r to deeper formations. However oil company lawyers overruled the engineers and said that the statute of limitations had run and that they would do nothing to help dispose of poisonous brines that I was pumping from the public aquifer.

What I learned from the episode was that early in the life of an oil well, the oil may freely flow to the surface because of the pressure that accumulated within the petroleum formation. As oil is extracted, the "bottom hole" pressure drops to the point that oil must be pumped to the surface. Corrosive oil brine increasingly comes to the surface along with the oil as the extraction process continues. The ratio of brine to oil increases until the cost of pumping the mixture approaches the market price of the crude oil.

In many ways the depletion of glacial age water is like oil depletion. Few would survive an Ice Age (even if we would foolishly attempt to arrange it) that might reestablish the once vast glacial-age underground aquifers. Similarly, it would tax human patience to wait for another 60 million years or more to see if we have another period of rampant fossilization of buried organic materials. Wasting either of these resources is extremely foolhardy.

These examples of the consequences of fresh water and petroleum depletion convinced me that it is increasingly imperative and urgently necessary to launch the Renewable Resources Revolution.

FIGURE 6.3

Oil brine disposal ponds, migration of brines into fresh water aquifer and wells with overlapping cones of depression to prevent brine trespass downstream

"Evaporation Pond"

Bad Water

Poisons

Contaminated Water

Good Water

Glacial Age Public Aquafer

Cone of Depression

The Common Defense:

Although I still fervently believe in free enterprise and government by the people, for the people, to provide the common defense, I am convinced that the common defenses that are most urgently needed are defensive actions against economic demise, environmental degradation and terrorists. Solutions to these looming problems are best provided by entrepreneurs who are less likely to be inclined to protect vested interests and more likely willing to take the business risks required to entice early-entry buyers to support needed products and services to facilitate the Renewable Resources Revolution.

WILL MONEY DO IT? Yes if the right thing is purchased.

Buckminister Fuller defined wealth in terms of the number of days that an individual would survive if he stopped being paid for doing work. Fuller reasoned that persons with large income from passive investments did not have to work or collect a paycheck to survive while less wealthy persons lived from pay day to pay day and few could survive more than a few weeks without a paycheck or some sort of welfare.

7.6

In fact, individuals once survived throughout an entire life without passive investments or employment contracts. Before money was invented, hunter-gatherers lived on the land. After farming practices were established cities developed and money was needed to provide a method for enticing farmers to grow food to feed the cities in exchange for goods and services provided by specialists in the city. Food, fiber, and leather takes a long time to grow and requires large risks by the farmer. Money enabled farmers to come to town and buy goods and services as desired through the seasons that would pass between harvests. Money also enabled merchants to travel for months to far away places and purchase goods that could be sold on their return.

Metals, particularly precious metals, offered compact forms of money that could be made into coins that allowed quick identification and agreement on value for desired purchases. Dollars were once guaranteed at $35 per ounce of gold. Extending this standard, silver certificates could be exchanged for silver on demand. Problems with this standard grew with population and the desire for world commerce. Money experts lobbied for rules that prohibited U.S. citizens from owning or accumulating gold.

In order to expand the amount of money to meet the increasing demands for borrowing which typically ratcheted up every year with the agricultural seasons, more gold had to be deposited in the federal coffers. Money creation was based on the amount of gold in reserve and each year more gold had to be bought, much of which was from speculators in Europe. After the big eastern banks that controlled the farm-area banks decided to signal for more money, it took at least 15 days to ship gold from Europe to the U.S. Until the gold was in reserve storage, banks would charge higher and higher interest rates for funds required by farmers to produce the food, fiber and leather for America. Farmers were increasingly forced into virtual peonage by high interest rate loans and many farms were lost to lenders.

Population increases and the desire to support foreign trade eventually made change necessary. In the 1960's the U.S. paid out billions in gold to foreign speculators and governments that collected dollars. And then the U.S. had to buy gold (much from the same speculators) to create money for the expanding economy and war in Viet Nam. By 1971 the U.S. owed foreign claimants demanding gold more than $36 billion but the gold inventory in Fort Knox and other depositories was only about $18 billion. The U.S. guarantee on the value of money was meaningless by virtual default.

President Richard Nixon was forced to "temporarily close the gold window" after depending upon it for funding most of the efforts to win WWII and rebuilding Japan, Germany, France, and Italy during the cold wars with the USSR and China. Richard Nixon had to announce that the U.S. could no longer pay out gold to buy U.S. dollars. By 1973, after exhausting efforts to find a way to continue with guarantees of monetary value but realizing that the value of the dollar was constantly decreasing, the U.S. conceded that the U.S. dollar would have to find its own value in world currency exchanges.

Instead of the gold standard that once provided world monetary stability for a small percentage of the world's population, the resulting sustainable energy and carbon standard will become a truly expanding value to back sufficient economic expansion to achieve our Solar Hydrogen Civilization's mission of worldwide prosperity.

Consequences were dramatic. Within six months after announcing that the dollar was floating, oil producers in the Organization of Petroleum Exporting Countries (OPEC) quadrupled the price of crude oil. They reasoned that the new "floating dollar" would be able to buy less. They expected that new dollars would be freely printed to pay trade deficits.

Today dollars are just special paper the supply of which is controlled in response to inflation and consumer confidence. If inflation threatens, the rate of supplying dollars is reduced and individuals hopefully become less willing to part with them for goods and services. In order to stimulate an economy, those given special privileges to run the monetary printing presses (the Federal Reserve) produce more money and get it into circulation through banks that loan it. Governments also may choose to stimulate an economy by printing checks to buy something. Illustratively, governments print checks to hire more employees, for welfare, or to buy something such as weapons, oil for sending navy fleets somewhere, jet-fuel to keep the strategic air command planes in the air, and other defense goods and services. Funds to cover the checks must be collected as taxes or borrowed.

FIXING THE DOLLAR: By buying the right thing.

7.7 Supplying government funds to hire more employees, pay for welfare, or for defense expenditures causes inflation because none of these efforts produces more goods. These efforts stimulate consumption of resources and cause inflation.

Another way a government could stimulate an economy and at the same time produce favorable anti-inflationary results is to buy storable supplies of renewable energy. Increasing the supply of energy helps homemakers, farmers, manufacturers, commodity transporters, and travelers. Increasing the supply of clean energy that actually enables existing and new engines to clean the air helps the environment and the economy.

Depleted natural gas and oil reservoirs and/or similar geological formations could be rented to store renewable methane and/or hydrogen. Existing natural gas pipelines can be utilized to transport the renewable methane and hydrogen to storage sites. Marketing the stored energy at a marginal increase will provide operating costs. Any person who demands to exchange dollars for valuable goods could chose to receive methane or hydrogen.

WHAT ABOUT THE DEFENSE INDUSTRY AND THE ARMED SERVICES?

Defense of the environment is imperative to protect our collective wealth from catastrophic events and our own inclinations to pollute and waste. Until required to do otherwise democratic and dictatorial societies have chosen to dump sewage and garbage in streams and into the oceans. Power plants dump gaseous pollutants, waste heat, and heavy metals such as mercury, lead and radioactive metals into the environment. Vehicle manufacturers produced a transportation sector that keeps cities on the list of dangerous places to breathe. Population increases have cancelled the option of diluting pollution to acceptable levels. Earth's oceans and atmosphere have been changed by pollution. Everyone is endangered for selecting the cheap options. Once a recommendation for good health, eating fish from contaminated water is now discouraged for children and pregnant mothers because of mercury accumulations.

7.8

Throughout the world, defense of the environment is needed. Restoration of civil, healthful, economic development conditions in response to every disaster is needed. Defense readiness and timely appropriate action is essential to overcome the challenges of catastrophic events including collision of our planet with a high inertia body in space, greenhouse gas changes to the global climate, communicable diseases, and to overcome terrorism.

Civilization needs 6,500 Renewable Resources Parks that are operated by non-profit organizations. These parks will be located in population districts with approximately one million persons and will take on the locally appropriate themes that combine primary energy conversion and waste disposal. Illustratively, Renewable Resources Parks will provide golf courses, nature walks, and botanical gardens along with solar, wind, wave, falling-water and biomass conversion systems.

NEW GI BILL TO STIMULATE SUSTAINABLE ECONOMIC PROSPERITY

A new "GI Bill" is proposed for soliciting persons from all branches of the armed Services to attend the International Renewable Resources Institute for purposes of receiving training and technology transfers to launch new ventures in their community of origin or other selected areas. This will help improve interest in joining America's all-volunteer military service by rewarding those who earn honorable discharges with opportunities to lead the country in transition to a wealth expansion economy. This will allow rapid development of new ventures throughout the country to advance each of the following essential economic developments to:

7.9

▲ Use renewable fuels and to sequester carbon for production of durable goods including longer-lived and more economical wind turbines, solar dish gensets, wave machines, and in-stream generators.

▲ Provide improved performance piping systems for collection and distribution of renewable energy, potable water, drip irrigation, and sanitary collection of wastes.

▲ Provide electrolyzers and hydrogen conversion kits for 800 million engines.

▲ Practice cogeneration along with electricity and methane/hydrogen wheeling.

▲ Provide improved roofing and siding materials for energy-efficient dwellings.

▲ Produce and utilize advanced engines, regenerative fuel cells, and recycled materials.

INTERNATIONAL RENEWABLE RESOURCES INSTITUTE
Launching businesses that make it convenient to participate in the revolution

One of the prevailing problems with traditional approaches for contributing to a better economy is illustrated by the plight of a person who aspires to be educated and graduates from some school with a diploma. The school, wanting to be respected and well supported by the community teaches what community leaders want and expect. Diploma in hand, the graduate seeks employment. After careful interviews establish that the person would be the best bet for fitting in and not rocking the boat, an invitation is offered to join the company team. An employment contract is established and the company shows the new-hire how to do what the company has been doing. Decades later the person retires having fit in as needed and the company continues to be dependent upon fossil fuels.

The result has been generations of remarkable progress that supports and enables a much larger human population than could be indulged without the multitude of technological advancements that characterize the last two centuries. But this progress blinds the fact that the Industrial Revolution has built a temporary economy that is rapidly depleting the beneficial environment and fossil fuels that supports the temporary economy that has developed.

The Institute will recruit the best and brightest entrepreneurs, engineers, technicians, accountants, advertising experts, etc., to prepare business plans for making, using, and/or selling goods and services that are based upon or that facilitate the advancement of renewable resources. The principal product of the Institute will be business launches by teams with complementary capabilities who prepare business plans that successfully compete in the global market for capital.

Operation of the Institute in conjunction with Renewable Resources Park(s) will enable hands-on learning where delegates from the world's most polluted cities and troubled economies will gain the knowledge, technology transfers, skills and business plans needed to prosper in various new business ventures to:

- ▲ safely produce, store, transport and use renewable fuels to provide energy-intensive goods and services;

- ▲ manufacture and market toys that inspire the emerging market to prefer everything from engines to homes that operate on hydrogen and clean the air;

- ▲ sequester carbon from sewage, garbage, and agricultural/forest wastes for production of durable goods;

- ▲ provide improved performance piping systems for distribution of potable water, renewable energy, and sanitary collection of wastes;

- ▲ provide sales and service of hydrogen conversion kits for 800 million existing engines ranging in size and types for applications in lawn-mowers to locomotives;

- ▲ supply, install, maintain, and operate fuel-cell and heat-engine cogeneration systems;

- ▲ produce, measure, and distribute renewable electricity;

- ▲ produce, measure, and distribute renewable methane/hydrogen by pipeline wheeling;

- ▲ provide improved roofing and siding materials for energy-efficient dwellings;

- ▲ produce, utilize, and market electrolyzers, hydrogen and oxygen storage systems, and fuel cells;

- ▲ produce high performance products from recycled materials;

- ▲ manufacture and market race cars, boats and airplanes that enable sporting interests to convince the emerging market to prefer higher performance hydrogen engines that clean the air and fuel cells that replace lead-acid batteries;

- ▲ produce water conserving drip irrigation systems that provide much longer life, improved farming productivity, higher crop efficiency, and improved economics.

What is needed to avoid the looming disaster due to fossil dependence is a departure from the "join the business and do the usual" that has been so successful to "lets launch new ventures to produce a wealth expansion economy." The International Renewable Resources Institute will define, refine, and demonstrate how Civilization can achieve sustainable prosperity by wealth-expansion practices in which an appropriate form of solar energy such as direct solar, wind, falling water, wave, or biomass is used to produce energy-intensive goods and services.

The non profit International Renewable Resources Institute will operate as a campus of sustainable technology learning centers and student operated technology demonstrations. Courses will be provided on production of hydrogen from sewage, garbage and farm wastes. Students will convert existing motor vehicles to operation on renewable fuels such as landfill methane, hydrogen, and Hy-Boost mixtures of these fuels. Renewable materials will be produced from biomass wastes and student/delegates will learn how to use existing components of the modern infrastructure to enhance economic development while achieving environmental protection. Students will qualify for graduation by producing business plans that successfully compete for investment capital and overcome traditional barriers to the use of renewable resources.

Subsequent chapters and the Appendix provide exemplary product descriptions that will be produced by new ventures that are launched by the IRRI. Selected technology transfers will support business teams that provide the products based upon technologies such as these shown in the patent disclosures listed in the Appendix.

Preventing Child Abuse:

Proposed sponsors of the companies launched by the Institute include organizations that have done so much to prevent deaths due to childhood diseases and now want these children to inherit a world that has good jobs with peace and prosperity. Rotary International, Lions International, Kiwanis, the Association of Energy Engineers, U.S. Department of Commerce, U.S. Department of Energy, and The Alliance to Save Energy are exemplary organizations. Profit motives will be emphasized in the development of new ventures to provide missing links such as conversion kits for heat engines, methane generation systems electrolyzers that are launched by the non profit Institute for facilitating the use of clean, renewable resources throughout the modern world.

An endowment of at least fifty million dollars is requested to sustainably launch the first phase of development of the initial campus of the International Renewable Resources Institute and to fund at least one Renewable Resources Park for energy production and hands-on education.

CAPITAL FLOW TO SUSTAINABLE COMMUNITIES:

7.12 Investment flows to projects that earn investor confidence. Communities that produce adequate supplies of renewable hydrogen to protect against fossil energy price escalations will be able to attract investment. Sequestered carbon products that offer improved performance will be efficiently produced by tooling that requires substantial investment to achieve desired economies of scale. Carbon products that are stronger than steel, lighter than aluminum, more corrosion resistant than stainless steel, and that can conduct more heat than copper will be produced.

SUMMARY OF KEY FACILITATORS OF THE RENEWABLE RESOURCES REVOLUTION:

As noted above, facilitating inventions will provide opportunities to develop much greater return on Civilization's enormous investments in infrastructure and equipment. These facilitating technologies will be commercialized by new ventures that are launched by the IRRI.

INVENTIONS: That enable:

Nearly any vehicle to be converted in the time of a tune-up to operation on hydrogen and/or and other renewable fuels.

Advanced storage systems that provide safer and more convenient hydrogen fuel storage than present methods of gasoline storage.

Production of hydrogen and carbon products from organic wastes. Carbon products ranging from super activated carbon to graphitic fibers that are ten times stronger than steel will be produced.

Electrolyzers that improve market conditions for renewable energy conversion projects and more economical engine/fuel-cell hybrids.

Hydrogen and carbon products from organic wastes.

Production of carbon-reinforced piping that is installed more quickly and that lasts much longer than conventional piping products.

Electronic contract metering systems that enable willing buyers to purchase renewable electricity, methane, or hydrogen from farmers and other entrepreneurs that supply these commodities by electric grid and pipeline infrastructures.

TESI: Total Energy Systems that provide emergency power for disaster areas and then support conversion of the recovering community to an economy based upon renewable energy and materials.

INTERNATIONAL RENEWABLE RESOURCES INSTITUTE: A non profit campus community that is self sufficient through the utilization of renewable resources. Hands-on training center for postgraduates and other particularly capable individuals to form new venture teams that provide products to produce a sustainable worldwide economy.

Cool!
Now that
it goes
faster
and cleans the air,
if it could just
give me a little
drinking water
from the
exhaust pipe.

THE SUSTAINABLE FUEL
HYDROGEN

AMERICAN HYDROGEN ASSOCIATION

SAVING THE WORLD

Lessons from history are disturbing. Two extremes are clear, either we develop renewable resources to produce sustainable prosperity or we discard decency and continue wasting technical prowess in ultimately self-defeating conflict over fossil resources. Other possibilities do not match the crisis caused by dependence upon burning over a million years' of fossil accumulations each year and human inclinations for self-defeating violence that are evident throughout history.

CHAPTER 8.0

EASTER ISLAND WARNING:

Easter Island is a tiny island in the vast South Pacific Ocean. It was named by European explorers who stopped there on Easter Sunday in 1722. It has a sad history that Civilization seems to insist on repeating on a global scale. Once it was a tropical paradise. Like all the people that traveled in ancient times to remote places, genetic evidence indicates that Easter Island was inhabited by settlers who derived from ancestors common to all humans. *Our Polynesian cousins sailed there in ancient times.*

It is startling that humans launched into the Pacific Ocean in vessels that would have violated all of today's safety codes. It is even more amazing that they found the tiny volcanic island in the vast uncharted Pacific by sailing at least 2,000 miles at an average of a few miles per hour. They had no clue that a hospitable place somewhere in the enormous Pacific Ocean could be discovered. Were these intrepid explorers risking everything because of a hostile send off? Were they sailing in faith of a mythical promise of paradise? Were they exceedingly foolish and/or amazingly lucky? How many other hapless expeditions were lost at sea before (and after) humans first discovered Easter Island?

Four formidable threats face human dominion on Earth. Failure to adequately prepare and overcome any of these threats could cause the same devastating end to human's 4 million-year dominion that befell the 160 million-year reign of dinosaurs. Three of these threats, namely collision with a high momentum mass from space, nuclear war, and catastrophic greenhouse gas accumulation, could cause mass destruction by severe climate changes. The fourth threat is economic demise and illness due to malnutrition and pollutant-compromised immune responses that fail the assault of naturally evolved and/or man made developments of communicable diseases.

After the very fortunate discovery, Polynesian settlements developed and were prosperous. Law and order took the form of interesting religious rules and regulations that leaders probably promulgated as ancestral myths were adapted to the island circumstances. Public works seemed to consist of carving increasingly large monolithic stone heads that were hauled to the coastline, apparently to watch for new arrivals or to protect the Islanders against some threat that was expected to come by sea.

Fishing was important to support Easter Island's rapidly increasing population. Island trees supplied building materials for housing, firewood, and most importantly for ocean-going fishing vessels. Trees that could provide timbers with sections needed to build sturdy vessels probably required 50 to 100 years to reach maturity. Humans have the capacity to double in population far faster and the Easter Islanders appeared to do so. Food production on the Island was not sufficient to feed the growing settlements that increasingly depended upon fishing. Easter Islanders found themselves to be dependent upon wooden ships to bring seafood to their growing population.

Demands for housing and fishing vessels to meet the needs of an increasing population brought strife as the Island's, once plentiful trees were depleted. As the Island's trees were depleted, hostile groups formed, vilified each other, and fought over remaining resources. Cannibalism emerged as the Island's cultural gains disappeared. When Europeans found the island in 1722, the Easter Islanders were represented by a few highly superstitious survivors with cannibalistic backgrounds who had lost most of the cultural gains and skills of their ancestors.

This example is a sad legacy of human nature. We have repeatedly developed technology to produce an economy, populated past the ability of the natural resources to sustain our needs, tried wealth redistribution in response to hardships that developed, and battled over remaining resources. Depletion of natural resources repeatedly drove humans to explore the far corners of the world's surface until fossil fuels were found in the underworld.

Exploitation of coal, oil, and natural gas allowed humans to rapidly demand six times larger food harvests and thousands of times more metal goods than any production level that could be sustained before the "Fossil Fuel Age." But depletion of the Earth's fossil reserves is inevitable. Even before depletion, hostile interests have developed concerning oil production rates versus growing demands for oil. Demand for energy exceeds production of depleting supplies of fossil and nuclear fuels. What are we doing about it? Much of the world is preparing for extended warfare by vilifying each other.

Human factions are fighting over fossil resources so the victor may have a few more decades of business as usual while advocating their religion or philosophy as the rationalization for beating the other side into submission. But it will be a futile victory that only delays the inevitable hardships that dependence upon fossil fuels has increasingly caused.

IS CIVILIZATION WORTH SAVING?
Sure, particularly if Civilization will save the world.

8.1 Enormous gains have been made in living standards by rapidly expending the Earth's fossil reserves. There is more to eat because of low cost energy for mechanized farming, fishing, and preservation options. We eat a much wider variety of foods because of energy-intensive canning, refrigeration and transportation. Tireless machines consume enormous amounts of energy to manufacture an amazing variety of transportation equipment, building products, and consumer goods. Ever advancing medical services are available to improve the quality and length of life. We have leisure time after meeting needs for food and shelter to conveniently travel quickly to distant places to visit interesting geology, cultures, or individuals.

The enormous gain that has been made in living standards is worth saving and it must be done quickly. In order to stop the waste of human resources, fossil fuels, and mineral resources required for hostile pursuits, it is imperative to take action immediately to accomplish a sustainable wealth-expansion economy.

Doing so requires much wiser use of remaining fossil resources. One gallon of oil can be made into $35 worth of durable goods. Renewable energy can be applied to extract about 5 pounds of carbon and 1 pound of hydrogen from one gallon of oil. The carbon can become machines that are tireless producers of renewable energy by harnessing solar, wind, wave, and falling water resources. The same approach can produce carbon and hydrogen from biomass wastes. Hydrogen can be used in existing engines to produce more power, extend engine life, and clean the air that passes through the engine.

Producing energy-intensive products from renewable energy resources including more and more machines that harness renewable energy will facilitate conversion of the fossil dependent Industrial Revolution into what Wall Street might ultimately refer to as first real Wealth Expansion Revolution or the Renewable Resources Revolution.

By converting to renewable hydrogen to fuel our activities we can reverse much of the damage that has been caused by burning fossil fuels. Damage being done by greenhouse gas accumulations will be reversed by carbon sequestration. Hydrogen produced from renewable resources along with hydrogen taken from fossil fuels (as carbon is sequestered) will replace hydrocarbon fuels. Sequestered carbon will be utilized to make lighter, stronger and more durable products instead of being added to the atmosphere by burning the fossil equivalent of 190 million barrels of oil each day.

Engines utilizing hydrogen will clean the air by converting greenhouse gas constituents such as unburned hydrocarbons, pollen, tire particles, and diesel soot into much less harmful water vapor and carbon dioxide. As shown in Table 3.1 a carbon dioxide molecule causes only a few percent as much infrared absorption as hydrocarbons such as methane, benzene, hexane, halogenated hydrocarbons and diesel-sourced aldehydes.

FIGURE 8.1

RESTORING NATURAL CONDITIONS IN THE STRATOSPHERE

Stratospheric ozone that protects life at the earth's surface from radiation damage and cancer due to high energy ultraviolet rays is being destroyed by chlorine and other halogens that have been delivered to the upper atmosphere by halogenated chemicals. We can remove stratospheric halogens and restore natural conditions.

Hydrogen and oxygen will be made by electrolysis of seawater. Large guns (revised naval guns) that utilize hydrogen and oxygen propellant or hydrogen and oxygen powered rockets will launch payloads of sodium and/or sodium hydroxide into the stratosphere. Atomized sodium orbiting the world in the upper atmosphere will react with halogens such as chlorine to form harmless salts that precipitate into the oceans.

How we will safely remove chlorine from the stratosphere.

ICBO Orbits Earth to Distribute Atomized Sodium
Na + Cl → NaCl

$$Na + Cl \rightarrow NaCl \qquad \text{EQUATION 8.1}$$

Figure 8.1 depicts placement of a very light weight radiation shield in the stratosphere to reflect intense solar radiation as it hosts reactions on the shaded side between sodium and halogens. This enables precipitation of sodium salts as chlorine bromine, iodine and fluorine are collected. These salts will eventually be returned to the oceans to eliminate the harm now done by halogens that destroy stratospheric ozone.

AVOIDING DOOMSDAY COLLISIONS

Large comets and asteroids that are on collision courses with our planet also threaten life on Earth. Such collision objects have pelted every planet in our solar system. Every day Earth collides with tons of small collision objects, most of which vaporize harmlessly in our atmosphere. In the past, collision objects large enough to cause substantial damage have hit the Earth. Many leading scientists believe that dinosaurs' 90 million year reign was ended 65 million years ago by climate changes that were produced by the high-energy impact of a collision object with Earth.

It is a matter of time until another asteroid will threaten life on Earth. An asteroid 15 to 30 miles in diameter could cause climate changes that wipe out Civilization if not virtually all life on Earth. Advancing our peaceful use of unmanned rockets provides a solution to a potentially life-

FIGURE 8.2

Disasteroids are coming, but when?

On January 18, 1991, an asteroid surprised astronomers as it suddenly passed 2.6 times closer than the Moon. Another close encounter came on May 20, 1993 when an asteroid narrowly missed the Earth by about the same distance. Another scary asteroid came 4 times closer than the Moon on December 9, 1994. The probability of being killed by an asteroid is about one in 5,000.

This is greater than the risk of being killed in a plane crash. This higher risk is because much or all of Civilization could be eliminated by collision with an asteroid less than 15 miles in diameter. More than 500 dangerous asteroids have orbits that cross the Earth's orbit around the Sun. However, about 300 of these potential colliders have been lost since they were first plotted. They are out of sight and out of mind because Civilization has decided that it is more important to have more people work at a single fast-food restaurant than to monitor the travels of potentially devastating asteroids and adequately prepare for diverting them from paths that could collide with Earth. The Solar Hydrogen Civilization must diligently prepare adequate preemptive ways to safely divert threatening "disasteroids" that are now hurtling through space.

threatening comet or asteroid. When a collision course is discovered, hydrogen-oxygen rockets will be launched to intercept the threatening object. A remotely controlled rocket will deliver a thruster engine to gently direct the "collision object" into a new non-threatening path. It will take much less energy to redirect a sizeable comet to a safe path rather than to blow it to bits and hope that none of the bits hit the Earth with life-threatening kinetic energy.

If needed, such objects may be guided into Earth orbit to allow mining of mineral values or utilization as water sources for production of hydrogen and oxygen. It will take far less energy to enter a water-rich object into Earth orbit than to launch water from Earth into orbit. Such a satellite could also host a greenhouse that converts solar energy into foods for future space explorers. This could serve as a fuel, oxygen and food-provisioning station for future space missions.

MENTORS: Saving the world starts with home, school, and neighborhood projects.

If you are good at teaching, designing, manufacturing, packaging, advertising, business planning, or financing important projects, please become a mentor at the International Renewable Resources Institute to help launch the businesses needed to facilitate transition to a wealth-expansion economy.

Or be the mentor who shows neighbors how to provide market pull for hydrogen and renewable carbon products. Some projects that you might consider follow.

THE HYDROGEN COOKING CENTER and BARBEQUE:

8.4 Hydrogen can readily replace carbonaceous fuels such as propane, charcoal and natural gas. Hydrogen burns about 7 times faster than common hydrocarbons such as natural gas and propane. Hydrogen diffuses and mixes with air much faster than hydrocarbons. If hydrogen is used to replace propane in a stove or barbecue burner, the fast flame propagation will probably cause backfiring in the tube that connects to the burner. However, because hydrogen mixes much faster than hydrocarbons it will probably burn efficiently and more controllably if it is not premixed with air before ignition.

Most of the fuel burners designed to use propane or natural gas can be converted to operation on hydrogen by:

1) modifying as needed to ensure that the cross sectional area of the fuel delivery passageways up to the final orifices is considerably larger than the total area of the orifices, and

2) blocking air entry into the fuel distributor.

These changes assure equivalent deliveries of hydrogen through all exit orifices and that only hydrogen (without air) passes through the tube that connects to the burner. If the orifices need to have reduced cross sectional area, several options are worth considering. One way is to plug the orifices with a braze metal that closes the orifice. Or a close fitting wire can be peened and swelled like a rivet to make it firmly anchored. A suitably smaller hole can be drilled through the manifold near the location of each plugged hole.

As shown schematically in Figure 8.3, this provides equalized hydrogen distribution through the orifices. Closure of the air passage into the carburetor tube eliminates the ability of flames to

FIGURE 8.3

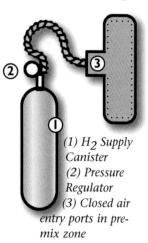

(1) H_2 Supply Canister
(2) Pressure Regulator
(3) Closed air entry ports in premix zone

enter the resulting "hydrogen-only" delivery system to the orifices. By providing a considerably smaller total area of the orifices than the cross section of the delivery system, equal flow of hydrogen will be maintained through all orifices and cooking can be accomplished without hot spots.

If you want the hydrogen flames to be more visible. the fuel manifold near the orifices can be coated with a solution of table salt and water, which dries to leave a residue of white salt. The burner gradually ingests some of the salt to cause the flames to be bright yellow-orange colored. The salt will be depleted in time and can be replenished. Salting the burner is advisable only if the burner is made of a corrosion resistant material such as stainless steel or if you quickly dry the saltwater solution and keep moisture from the salt deposit to prevent formation of a corrosive electrolyte.

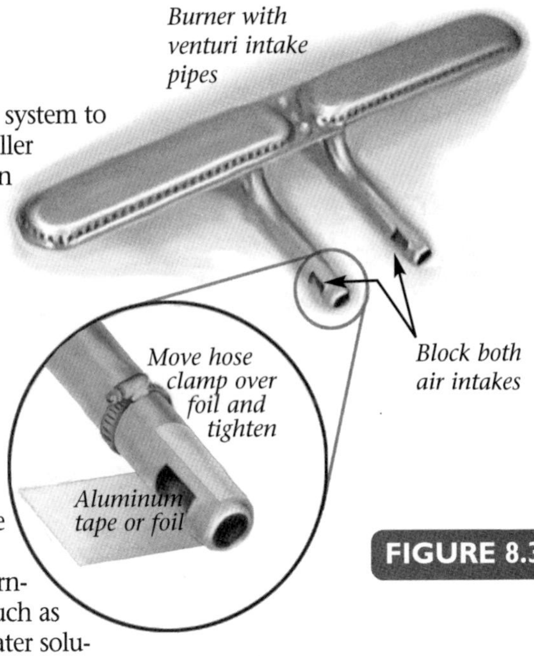

Move hose clamp over foil and tighten

Aluminum tape or foil

Block both air intakes

FIGURE 8.3.1

Blocking air entry into the manifold is best accomplished by making a closely fitting sleeve that covers and seals the air passageways. If the air vents are not sealed air will enter or hydrogen will leak through them. In either event flames may result in the wrong places. In many instances air inlet vents are made by machining holes in a cylindrical tube. This type of vent can be sealed by several excellent methods. A sleeve of heavy aluminum foil that is wrapped several times around the air vent and held with two clamps that seal the area has been found to be a quick solution. Another method is to cover the area with an aluminum-covered fluoropolymer shrink tube.

A compressed gas storage tank such as might be used for compressed air in scuba diving or for storing natural gas can be re-rated for storage of hydrogen. The tank manufacturer should be contacted for details about converting it for hydrogen storage. In many instances the tank will already have manufacturer's approval for safe storage of hydrogen. If not, encourage the manufacturer to offer products with specific approval for hydrogen storage.

Using the tank for hydrogen storage requires flushing out any oxidant, fitting it with a thermally activated pressure relief device and providing it with an approved fitting for safe and swift refilling operations.

HYDROGEN ENGINES CLEAN THE AIR:

Start with a small engine that can be conveniently operated as an air cleaner. Measure the emissions from gasoline operation for reference. **8.5**

Small engines usually pollute much more air than larger engines that have been equipped with emission controls. Small engines are typically allowed to emit carbon monoxide, hydrocarbons, and oxides of nitrogen without after treatment by catalytic reactors. A small engine may produce 5 to 20 times more pollutants per hour of operation than a modern automobile engine. Converting the conventional gasoline engine used for garden equipment, go-carts, engine-generators, and motor-bikes can overcome pollution equivalent to 5 to 20 neighborhood cars.

So, if you mow a lawn or provide a lawn-mowing service, convert your lawnmower to hydrogen. Engines of all kinds can be converted to operate on hydrogen. Several methods work well.

First clean and check the condition of the engine for achievement of manufacturer's specified performance. Inspect the spark plug and "read" the engine condition as noted in the manufacturer's repair and maintenance manual. Check the compression pressure. Remove the head and inspect the combustion chamber to remove any carbon deposits. Repair the engine if necessary according to the manufacturer's recommendations and specifications.

In any event remove excess carbon that has been plated in the combustion chamber or on the valves. Carbon deposits may glow hot enough when the engine is running to cause ignition of hydrogen before the intended spark ignition. Carbon deposits that were previously cooled by evaporation of gasoline will run hotter; possibly hot enough to cause backfiring if hydrogen is added as a homogenous charge with incoming air.

Small four-stroke engines with magneto ignition can be converted by providing a metering valve that delivers controlled flows of hydrogen to the intake valve port. When the intake valve opens, a mixture of air and hydrogen flows into the engine. It is preferred to block the throttle valve wide open to prevent the engine from producing a vacuum in the intake manifold. Ignition timing may need to be set closer to top dead center than with gasoline. The best ignition timing is found experimentally by adjusting the point of ignition until the highest engine speed is produced for a given fuel flow.

An engine with a flywheel magneto-ignition system can be operated with experimental ignition timing by removing the key that locks the position of the flywheel to the crankshaft. Adjust the timing and re-key to permanently provide the optimum timing after it is experimentally found.

An improvement on the constant flow of fuel into the intake valve port consists of a timed port injection control valve that meters hydrogen into the combustion chamber each time the intake valve opens. This approach is easiest on engine applications that have a battery for operation of a starter motor. The battery can be utilized to power a normally-closed solenoid-operated valve that is timed to open each time the intake valve opens. This avoids accumulation of hydrogen in the intake manifold between intake

FIGURE 8.4

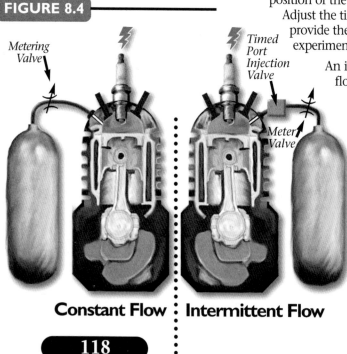

Metering Valve

Timed Port Injection Valve

Meter Valve

Constant Flow : **Intermittent Flow**

Ambient Atmosphere Of Test Shed	28 ppm HC (airborne hydrocarbons)	0.00 (carbon monoxide)	1.00 ppm (oxides of nitrogen)
H2 Idle	19 ppm (eliminated of 9 ppm hydrocarbons)	0.00 (no carbon monoxide)	1.00 ppm (added 0.00 ppm oxides of nitrogen)
H2 Full Power	8 ppm (eliminated of 20 ppm hydrocarbons)	0.00 (no carbon monoxide)	1.7 ppm (added 0.7 ppm oxides of nitrogen)
Gasoline Idle	180 ppm (added 152 ppm hydrocarbons)	24,300 ppm (added 24,300 ppm carbon monoxide)	410.00 ppm (added 409 ppm oxides of nitrogen)
Full Power On Gasoline	192 ppm (added 164 ppm hydrocarbons)	8,100 ppm (added 8,100 ppm carbon monoxide)	110 ppm (added 109 ppm oxides of nitrogen)

NOTES: *ppm = parts per million concentration*
both tests without catalytic reactor in exhaust system
gasoline engine operated according to manufacturers specifications
hydrogen engine operated without air throttling

valve openings and the tendency of constant flow systems to occasionally have fuel-rich combustion conditions.

Figure 8.4 shows two types of homogeneous-charge operation. A constant flow system is compared to an intermittent flow system for small engines. Table 8.1 shows comparisons of "shed tests" in which an engine is operated in air that has a known concentration of airborne hydrocarbons. Hydrogen fueled engines can clean the air that enters their combustion chambers.

TABLE 8.1

Emissions test comparison with gasoline.

CONQUERING DOOMSDAY GERMS
while doubling energy utilization efficiency.

An important way to help escape doomsday diseases is to clean the air that you choose to breathe by first passing the air **8.6** through the combustion chambers of a hydrogen engine. The air will be suddenly steam sterilized at 4,000°F (2,200°C) to convert prions, viruses, bacteria, pollen, anthrax and mold spores along with other potentially dangerous germs into traces of harmless water vapor and carbon dioxide. Exhaust from the air-cleaning engine should then be passed through a sterilizing bath of boiling water to further assure safety.

The amount of water vapor and carbon dioxide that will be produced when a life-threatening inventory of the most dangerous germs are dissociated by 4,000°F steam is miniscule. Up to about 70% of the substance of pathogenic germs is composed of water. If the combined amount of bio-terror germs that could cause respiratory anthrax, small pox, and measles were dissociated at 4,000°F in the combustion chamber of an engine, the amount of water and carbon dioxide produced would be less than the weight of a few grains of table salt. This is a miniscule amount compared to the nine pounds of water that the engine will produce by combustion of each pound of hydrogen. Carbon dioxide produced by hydrogen induced burning such an infectious inventory of bio-terror germs will be many orders of magnitude less than the 20 pounds of carbon dioxide now produced when an engine burns a gallon of gasoline.

FIGURE 8.5

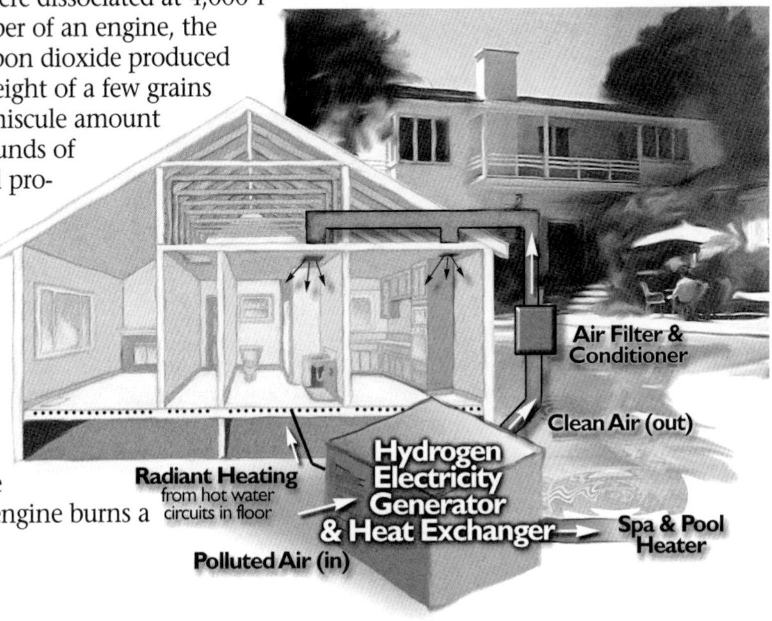

Air Filter & Conditioner

Clean Air (out)

Hydrogen Electricity Generator & Heat Exchanger

Radiant Heating from hot water circuits in floor

Spa & Pool Heater

Polluted Air (in)

800 MILLION ENGINES CAN CLEAN THE AIR: and make a market for producing 1000 new millionaires each week.

8.7 VIRTUALLY ANY ENGINE can be converted to operation on hydrogen in the time of a tune up with the devices shown in Figure 8.6. Converted engines can clean the air. Because there is no carbon or sulfur in the fuel, engines do not suffer from corrosion by acids that are formed from oxides of carbon and sulfur. No acid rain sourcing emissions are produced. Abrasive wear is reduced because there are no particulates from hydrogen combustion. Engines last longer, produce more power, and clean the air.

FIGURE 8.6

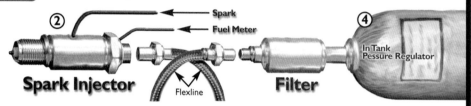

CONVERSION: Spark plugs or diesel injectors are removed and replaced with the Spark Injector devices (2) as shown. Hydrogen storage tank (4) supplies hydrogen at regulated pressure to the Spark Injectors that have built in microcomputers that sense the piston position and combustion chamber events to optimize performance.

OPERATION: When the piston approaches top dead center on the compression stroke, the Spark Injector injects hydrogen to form a stratified charge that is ignited by one or more sparks that are produced in accordance with an adaptive timing that is controlled by the built-in microcomputer. Engines start in the coldest weather and do not hesitate in the hottest climate as facilitated by direct injection and spark-ignition of hydrogen in the combustion chambers as the power stroke begins. Fuel economy and power production capacity are improved.

To assure continuity and availability during transition to the hydrogen economy, Spark Injectors allow any engine to be interchangeably operated on hydrogen or on gasoline. The gasoline system can be retained exactly as provided by the original manufacturer except the spark plugs are replaced by Spark Injectors. In case an operator chooses gasoline, the Spark Injector simply provides the spark at the same time it is ordinarily provided for gasoline operation.

A much more fuel-efficient way to use gasoline is to operate the gasoline system with adjustments that provide a gasoline-air mixture that is too lean for spark ignition. Ignition is provided by timed injection of hydrogen to produce stratified-charge combustion of hydrogen within the very lean gasoline-air mixture. This provides assured ignition and much faster combustion of the gasoline for improved fuel economy and performance. Emissions are greatly reduced compared to normal gasoline operation.

The best way to operate the vehicle is to shut off the gasoline system and operate on directly injected hydrogen. In this mode it is possible to improve volumetric efficiency of the engine by allowing un-throttled air to enter the engine at all times. Fuel economy and performance improvements are substantial. Engines last longer and clean the air.

Everyone is a Winner:

800 million engines are candidates for hydrogen conversion to air cleaning status. Hydrogen Corridors and cities with air pollution problems can produce clean air and foster economic development by providing incentives for conversion of existing vehicles. The Clean Air Lottery is proposed as a way for everyone to be a winner. *Everyone that breathes, drives or benefits from economic development is a winner.*

The clean Air Lottery provides much higher returns to the gambling public than any other lottery in history. How it makes everyone a winner follows:

1. Each purchase of gasoline will include a dime per gallon chance on winning the Clean Air Lottery.
2. Every day a group of free installations for "conversion kit" winners will be announced until all the registered vehicles in the lottery district are converted.

BIGGEST GAMBLE IN HISTORY
can be converted to the *solution to pollution*.

People who buy and burn fossil fuels are creating the biggest gamble in history. We are gambling with the global climate and the economic welfare of an enormous population that has been produced by betting on fossil fuels. Every gasoline customer is already taking a chance with each purchase on lung disease, greenhouse gas induced weather damages, MTBE water pollution, and ultimate conflict over remaining oil supplies in order to continue the global petrol habit.

The clean air lottery provides much higher returns to the gambling public than any other lottery in history

The CLEAN AIR LOTTERY will utilize a computer program to award certificates for a free installation of hydrogen conversion kits. Each week a grand-prize winner of one year's supply of hydrogen will be provided to the lucky person who is selected from the week's winners of conversion kit installations.

8.8

Municipalities that dispose of garbage, sewage, or farm wastes can produce carbon products and hydrogen from these wastes and assuredly provide hydrogen at a lower cost than gasoline. Carbon products will be provided from industrial parks that produce lighter, stronger and better carbon products ranging from sporting goods to corrosion resistant products for marine environments.

As the demand for hydrogen grows, other production sources can be bond-financed to continue the sustainable economic development that has been launched. Positive economic development will flourish as dollars formerly sent out of the economy for petrol and health care circulate through local businesses.

Of course, oil companies will be welcome to take carbon from gasoline and sell hydrogen at prices established by renewable hydrogen production from wastes. The emerging Carbon Revolution needs the carbon for much more important utilization in durable goods than having it burned to produce pollution and economic inflation.

Proposed sponsors and promoters of the Clean Air Lottery in your neighborhood include non-profit organizations that seek local solutions for local problems. The Clean Air Lottery makes a much larger return to each petrol-dependency gambler than the amount gambled.

Other potential sponsors include:

1. Health-oriented organizations who recognize the great value of preventing illness by replacing polluted air and water with clean air and water.

2. Racing and performance enthusiasts want hydrogen to enable existing engines to produce more power and last longer.

3. Distributed energy promoters who realize what hydrogen fueled energy conversion regimes greatly improve operating efficiency and reduce maintenance costs.

4. Economic development promoters who realize the great benefits that follow production of hydrogen from local renewable resources such as solar, wind, wave, falling water, and biomass wastes. Reversing the trade imbalances due to fossil fuel importation to a region enables much larger investment in local economic-development pursuits.

5. Tourism promoters who realize that travel will increase to areas that provide clean air and water benefits along with energy-security and peace of mind to travelers.

6. Governmental authorities who now struggle unsuccessfully with smog, water pollution, and jobs losses.

7. National authorities who are committed to economic development and energy security. U.S. oil imports now cost over $1 billion every three days. This means that sustainable economic development "equivalent to 1,000 new millionaires" is needed each week to produce and distribute renewable hydrogen and methane to earn the $2 billion that is now spent on fuels that pollute the cities and poison community water supplies.

All this from Clean Air Lotteries that make everyone a winner!

**ANOTHER LOTTERY FOR
AVOIDING THE EASTER ISLANDERS' FATE:**

Easter Islanders allowed themselves to create problems of overpopulation, economic inflation, depletion, and hardship because of poor leadership. Island leadership obviously followed so many previous examples of resource depletion and warfare over remaining supplies. This sad syndrome is being repeated by the U.S. and other industrialized countries that now continue to be dependent upon diminishing reserves of oil and other fossil fuels.

8.9

Easter Island leaders must have had the same human inclinations to resist change. Their leaders must have reasoned that they would be on the winning side of wars over the diminishing supplies of timber and farming areas. Victory that would provide temporary control over scarce resources was perceived to be more favorable than to risk giving up institutionalized habits that supported the leaders' importance and power.

Generally, political leadership of industrialized countries is following the same well-established road to ruin. Burning the Earth's finite hydrocarbon reserves at the rate that is equivalent to a million years' of fossil accumulations per year is unconscionable… unless it is done with clearly defined goals to do so in support of a swift and efficient transition to an economy based on renewable resources to produce sustainable wealth expansion.

Term limits, reduced election expense, and more efficient government operations:

It seems logical therefore in this time of remarkably expensive elections and highly lobbied incumbent politicians that local leadership might be better found by another type of lottery… the Leadership Lottery. Winning a one-year, incentive based opportunity to double the average voter's wage, by serving in a public service appointment would result from a randomized computer selection from qualified residents of a community.

Qualification of candidates would be established by solicitation of written proposals that would be published in local newspapers. The ten proposals that receive the greatest support in opinion polls would enable their authors to be "qualified" for random selection in the double the average wage lottery.

Another qualification requirement of candidates would be prior agreement that all proposals and suggestions would be published and made available to the lottery winner for purpose of public benefit.

A key to success for the Leadership Lottery is to provide incentives for substantially increasing the income of lottery appointed persons by reducing the cost and increasing the public benefits provided during leadership tenures.

Accomplishment of clearly stated goals that are proposed by qualified candidates would be rewarded with a bonus at the appointment conclusion. A bonus magnitude that doubles the average annual income of taxpayers seems appropriate. With such an incentive it will be more difficult for lobbyists to buy control of one-year appointees with clear plans and incentives for reducing government waste and excesses. Short-term appointees will hopefully provide more opportunities for good ideas and diligent attention to governmental efficiency than provided by present political systems.

Some of the first needs for such a Leadership Lottery could provide appointment of government office holders that control waste disposal systems and energy purchases. Written letter proposals for improving the performance and efficiency of waste disposal operations would be published and editorial comments and criticism would be invited. Subsequently, a randomized computer selection would award the contract for employment to the person selected from the ten most desirable proposals from qualified residents as indicated by an opinion poll.

Much improved opportunities for economic development could be achieved by conversion of municipal wastes into hydrogen and carbon. Savings in electricity costs could be provided by installation of renewable energy co-generation systems to replace electricity from fossil fueled central power plants. Government carpools and equipment fleets could be converted to hydrogen operation. Substantial reduction of air and water pollution will be collateral benefits. Public buildings including schools could be provided with distributed energy conversion systems that double energy-utilization efficiency while providing sterilized air to overcome bio-terrorism threats.

OVERCOMING CHILD ABUSE AND TERRORISM

In spite of the oil revolution and ever-increasing energy consumption that brings enviable affluence to industrialized nations, a far larger portion of the world's population has too little disposable energy to enable heads of families to earn enough to keep their children from suffering child abuse by starvation and/or crippling hardships from poverty. Civilization needs more energy than all of the Earth's oil, natural gas, coal and nuclear fuels can supply.

The Solar Hydrogen Civilization will provide the energy needed and overcome such child abuse by developing honorable jobs for more than two times the combined automotive and fossil energy sectors of the present global economy. It is imperative to do so with swift resolve.

Transition to the Solar Hydrogen Civilization needs to be done with global awareness, reason, fervor, compassion, and dedication to the cause of Sustainable Prosperity. Because the petroleum-supply circumstance will become more critical and because of the large and growing global population that is at stake, this cause must be known and endorsed by far more humans than the cause of any previous revolution.

It is a revolution that must invite participation by all age groups. It must involve children who are soon to inherit the Earth, for children they will create, and it must be done in ways that end child abuse and terrorism.

CHILD DEVELOPMENT
Civilization's Dearest Investment:

Unquestionably, one of the most basic investments that Civilization makes is the creation and nurturing of children. We expect and have great needs to pass the baton of progress to succeeding generations. We place our hopes on their success. Parents, extended family members, communities, and nations do everything possible to create better opportunities for children to be cared for, educated, trained, and launched to become successful adults. As evidenced throughout recorded history, parents have taken any necessary risk, diligently worked to support, and when needed sacrificed as much as possible to improve children's chances of success.

Doctors, nurses, and other practioners of midwifery specialize in prenatal care, planning, and assistance to provide more successful infant birthing. Specialists in childcare and development look after every stage of childhood. Teachers train to provide the learning atmosphere and stimulation that enables children to know much of what is needed to succeed in subsequent stages of education and advancement to adulthood.

CHAPTER 9.0

In addition to investing greatly in monetary and emotional ways in beginning stages of life, Civilization supports vocational schools, colleges, and universities that formulate programs for further specialization and educational discovery. One third of a person's life is often devoted to learning about what might be done for his or her remaining time on Earth in fulfillment of professional goals. In some traditions this learning experience includes an apprentice period of concentrated technology transfers. In other traditions that embrace technological advancement, the educational process is best dispersed throughout the career of professional practice.

GROWING CHILD ABUSE:

Victims of child abuse and terrorism share the common experience of being injured by persons who disguise their intent in order to cause harm. Even the most casual observations of child development reveal that children do what their parents do until more powerful examples are provided. Watch little girls emulate their mothers in caring for baby dolls. Watch little boys try to duplicate and use tools that their fathers use. If a dominant member of a family abuses other family members or others in the community, expect the same behavior to be advocated if not repeated by children. If a community or nation hosts radical leaders who advocate terrorism, children will be far more likely to accept or become supporters if not participants in terrorism.

9.1

Child abuse and terrorism are serious detriments to the gains that have been made by Civilization. Child abuse harms childhood progress and in many ways disadvantages the future of abused persons and of persons who depend upon them. Terrorism seeks to disrupt and distract if not defeat orderly pursuits of freedom and accomplishment of higher living standards. Terrorism and counter-terrorism measures reduce freedom for every family and community on Earth. More children than soldiers are killed or maimed by land mines left behind as a result of wars.

Child abuse takes many forms. Undoubtedly the most pervasive forms of child abuse are various manifestations of malnutrition and diseases that are enabled by weakened immune systems due to economic poverty and environmental pollution. Parents who bring children into such hopeless economic hardships are responsible for this type of long-term abuse. Institutions that encourage parents to participate in this kind of child abuse are equally responsible.

Related forms of child abuse stem from widespread environmental pollution. Dependence upon fossil fuels has caused rapid increases in the presence of environmental contamination by petrocarbon particulates that cause lung disfunctions and by heavy metals such as lead, mercury and radioactive metals. Child development and accomplishment of genetic potentials are sacrificed as a result.

A study by the Harvard School of Public Health showed how fine-particle emissions from nine coal-fueled power plants in Illinois were linked to an estimated 300 deaths, 13,900 asthma attacks, 2,600 emergency room visits, and 500,000 incidents of upper respiratory disease each year. In many other areas and particularly in developing countries, emissions from fossil fuel burning power plants are more noxious than from the coal-fueled plants in Illinois. Children who succumb to particulate caused lung diseases are innocent victims of adult insistence upon continuing dependence upon fossil substance abuse.

Similar links between burning fuels and disease exist in every industrialized area of the world. Pollution including fine-particles travels from industrialized areas throughout the world. Coal plant pollution produced in China crosses the Pacific and causes air pollution in U.S. cities. As will be noted, many jobs, more than any war would provide will be needed to produce renewable hydrogen and electricity to overcome illnesses caused by burning fossil fuels. The question remains: when will we overcome the multiple dilemmas of economic hardship, diseases that handicap children and hamper every age group, and conflicts over finite resources that dependence upon fossil fuels causes?

Children with little hope of gaining the advantages of educational training to win good jobs and community respect often experience emotional distress. Increasingly, children face a life of futility, frustration, and exploitation. As in most instances of surplus, the value of individuals who constitute the surplus is low in the eyes of leaders of such populations.

Illustratively, children in North Korea, Iraq, and many African countries are subjected to inhumane treatment for political purposes that too often coincide with personal gains of government insiders. Relief funds and contributions of equipment, food and medicines are often used to secure weapons instead of being distributed to impoverished families. Children suffering from malnutrition and polluted water die from diarrhea and pneumonia. For more than a decade, it is reported that 250 malnourished children died each day in Iraq, but Saddam Hussein used income from oil sales for his opulent palaces and to buy weapons from North Korea and from arms dealers in many other countries.

North Korea is intent on maintaining a large standing army and on becoming a provider of long-range rockets and nuclear weapons. After decades of hardship, North Korea remains under control by the regime of KIM Chong-il who manages to intimidate the world community sufficiently to be able to rely heavily on international food contributions to prop-up his hungry army and starving nation. North Korea maintains a standing army of about one million battle ready troops and continues to expand a long-range missile development program along with efforts to produce nuclear, biological, and chemical weapons.

Former agriculture diplomat Kim Dong-Su defected from North Korea and gave a chilling account of North Korea's three-year famine and growing economic crisis. Kim Dong-Su's defection followed executions of dozens of senior party officials who voiced complaints about the plight of North Korea. One of the officials who KIM Chong-il executed was Kim Dong-Su's former boss, Suh Kwan Hee, who was the communist party agriculture secretary. Suh Kwan Hee reportedly voiced a protest against North Korea's military priorities that deprived needy families of relief and caused civil discontent.

According to Kim Dong-Su, up to 2.8 million people may have died from starvation while communist leaders directed public wrath at the U.S. as the scapegoat for the economic hardships of North Korea. Grain and other relief supplies were confiscated by military authorities while the civilian population suffered. Youth who joined the military received food, clothing, weapons, and training to protect the militant communist regime of ruler KIM Chong-il.

According to United Nations reports regarding famine in North Korea, one child in seven suffers from acute malnutrition, and citizens throughout the country face slow starvation on a massive scale. Following massive floods that devastated its persistently under-performing food production system, North Korea received the largest U.N. relief operation in history. Some 650,000 tons of grain were donated by the international community along with medicines and various other supplies to support 7.5 million of North Korea's 22 million population. However, North Korea spends funds, including relief funds, on military pursuits instead of providing food and clothing for impoverished families and starving children.

Ruthless leaders obviously find it preferable to sacrifice the time and energy of surplus young people in military pretentiousness and hostile acts to gain political outcomes and to confiscate resources. Forcing hungry youth into military obedience is less objectionable than to face insubordinate civilian youth who foment discontent, criticism, and possible revolution. According to United Nations authorities and as noted in "Child Soldiers Global Report 2001," more than 300,000 children under 18 are soldiers in armed conflicts in more than 30 countries around the world.

Recognized governments and armed resistance groups exploit children because they are easier to condition for unthinking obedience and high-risk assignments. Land mines are often placed or cleared by young children. Most child soldiers are between 15 and 18. Even younger children have been trained to spy, carry messages, and hand carry remotely controlled explosives to enemy shopping centers and strategic locations.

Military leaders often accept rape, sexual harassment, abuse, and sexual slavery of young girls. Sexual slavery produces children who are born into a life of disdain and exploitation that perpetuates the malicious syndrome of child abuse. Children abusing children is a progression of this sad syndrome.

GROWING TERRORISM

9.2 Street gangs extort cooperation by cruel acts against those that resist. Terrorists such as kidnappers prey upon weaker or vulnerable persons because they can use surprise, access, or some other situation to advantage. Kidnappers often seek extraction of some kind of ransom or other consideration such as trade of the kidnapped person for an imprisoned person. Another reason for terrorist attacks, is to avenge some perceived wrong that originated from, is supported by, or condoned by identified communities with opposing beliefs.

Mental conditioning of persons who commit terrorist acts takes many forms and ranges from risk/reward programming to exploitation by assertive leaders who gain authority to do so by self appointments and/or by duping established institutions into exploitative training. Corruption, political oppression, poverty, ignorance, disease, and environmental disorder are interactive harms. These harms have been cited by sociologists through the years as circumstances that cause violence, civil disobedience, and child abuse.

Exploitation of one or combinations of these hardships enabled ruthless dictators such as Napoleon Bonaparte, Joseph Stalin, Adolph Hitler, and Benito Mussolini to gain the assistance of military and industrial interests to do great harm to the interests of fellow citizens and then to threaten the world with militant totalitarianism. More recently terrorists have emerged from camps of religious tyrants who blame Western culture and capitalism as evils that must be combated by Al-Qaeda or Taliban warriors. Western culture is blamed for poverty in dictatorships that ignore deplorable living standards.

Ironically, Osama bin Laden is a Saudi multi-millionaire who received U.S. backing in the war anti-Communist Afghan guerrillas fought with the Soviet backed government of Afghanistan from 1979-1989. A decade later, Osama bin Laden recruited privileged Saudis and Egyptians to hijack and crash four civilian airliners. Two were crashed into the World Trade Center Towers and one hit the Pentagon on September 11, 2001. What was the fourth target?

LET'S ROLL:

9.3 The fourth flight that was taken over by hijackers was United Flight 93. Forty one of the forty four persons who boarded Flight 93 in Newark N.J. on September 11, 2001, expected to land about five hours later in San Francisco, California. However, three passengers were Al-Qaeda terrorists who demonstrated their brutal intentions by stabbing another passenger to death. Shocked passengers and cabin crew members were forced to the back of the plane. Demanding control of the aircraft on threat of killing more passengers, the terrorists commandeered United Flight 93 and took control of the cockpit. Flight 93 traveled to the Cleveland area before diverting towards Washington D.C. But what was Al-Qaeda's intended target?

Several of the passengers managed to have cell phone conversations that alerted them about the two previous hijacked planes that had been crashed into the twin towers of the World Trade Center. Realizing their plight, Todd Beamer, Elizabeth Wainio, Thomas Burnett, Jr., Mark Bingham, Debby Welsh, Toshiya Kuge, CeeCee Lyles, Jeremy Glick, Linda Gronlund, and other passengers decided that they would refuse to ride complacently to their death. But what could they do as the terrorists continued with plans to use Flight 93 as a 500 mph jet-fuel bomb to kill more victims at their next target? CeeCee Lyles was a flight attendant who managed to use her cell phone to call her husband and tell him that she loved him and their children. Other cell phone conversations included the similar sentiments as a plan evolved.

One terrorist had what he claimed to be a bomb strapped to his body. He insisted that passengers stay seated in the back of the plane or he would detonate the bomb. After being alerted to the fate of persons on earlier flights that had crashed into the World Trade Center Towers, passengers and flight-crew members decided to try to overcome the terrorist with the bomb that guarded them.

If the bomb went off and blew the plane up it would save lives of innocent victims at the unknown target that Flight 93 was speeding to destroy. If they could disable the guard who had herded them to the back of the plane, they would form into a single file in the narrow isle to the cockpit and keep charging ahead with their combined momentum until someone broke into the cockpit to evict the two terrorists who had control of the Boeing 757. Todd Beamer, a 32-year-old businessman, Sunday school teacher, husband, and father was heard on a cellular telephone reciting the Lord's Prayer. Beamer finished the prayer and gave the command to start their charge saying "Let's Roll!" They succeeded in overcoming their guard and some of them held him down.

Pushing a service cart as a battering ram, the remaining task force was able to bash the cockpit door open. The terrorists in the cockpit apparently chose to crash the plane instead of allowing the passengers to take control and save those on Flight 93.

GREED OR RELIGIOUS INTOLERANCE?

In 1980, Iraq, with the world's 2nd largest oil reserves attacked Iran the world's 5th largest oil holder. Iraq supposedly did so to control a historically disputed area called the Shatt al Arab waterway that separates Iran and Iraq. Iraq continued with military offensives, however, in an effort to seize the western Iranian region known to have substantial oil reserves.

9.4

Ayatollah Ruhollah Khomeini, the riled Iranian spiritual leader of Iran, declared that Iran would not stop fighting until the regime of Saddam Hussein was toppled. Military training and indoctrination was quickly extended to schoolchildren. Iran urgently bought arms, increased army ranks, particularly with teenage boys under 18, and mounted counter-offensives that incurred severe casualties. Iran's massive counter-offensives managed to force Iraqi forces to leave captured areas of Iran. Aircraft and missile attacks were increasingly exchanged on the capitals of Iran and Iraq.

Both Iraq and Iran are parties to the Geneva Protocol that prohibits the use of asphyxiating, poisonous and other gases, and all analogous liquids, materials or devices, as well as the use of bacteriological methods of warfare. However, Iraq launched chemical weapons attacks on Iranian troops and civilian populations. In 1980 Tehran Radio stated that Iraqi air strikes on Susangerd caused deaths by chemical-gas bombs. Of the 49 reported instances of Iraqi chemical warfare attacks in 40 border regions, several unexploded chemical weapons were recovered, investigated and the contents verified by inspection teams of the United Nations.

Poisonous gas samples taken from unexploded bombs included mustard gas and tabun nerve gas. Mustard gas is typically delivered as an oily liquid that evaporates slowly to remain a poisonous threat for days or weeks. Taken into the body it is a systemic poison that is deadlier than hydrogen cyanide. Mustard gas acts more slowly however, than cyanide to cause poisoning. Some hours after exposure to mustard gas, symptoms such as blindness, blistering and lung burns are caused. These symptoms were noted in patients sent from Iran to hospitals in Switzerland, Sweden, Netherlands, Japan, Germany, France, Great Britain, Belgium, and Austria.

Tabun nerve gas is also delivered as a pressurized liquid in a bomb, artillery or mortar shell, or land-mine canister. Tabun evaporates even more slowly than mustard gas. Tabun stays where it is released much longer and it acts much faster than mustard gas. Tabun was used to stop Iranian infantry assaults. Air-burst tabun bombs could be dropped by warplanes that strafed and poisoned enemy positions.

Sarin and Tabun were secret chemical weapons produced by Germany in WWII. Germany manufactured 12,000 tons of Tabun during 1943-1944. Tabun samples taken by United Nations investigators from victims of Iraqi attacks indicated that the Iraqi tabun's "chemical fingerprint" was the same as from tabun produced by methods developed in Germany during WWII. Tabun inhalation causes involuntary urination and defecation, vomiting, twitching, convulsions, paralysis, unconsciousness, and death.

Victims of Iraqi attacks also were diagnosed to have died from anthrax and undiagnosed chemical agents or mycotoxins that left no trace of tissue injury. In some attacks, corpses of Iranian civilian and military personnel were examined who "looked as if they had fallen asleep." Similar deaths were reported after Iraqi attacks on Kurdish villages in Iraq. Iran was desperate to stop the flow to Iraq of weapons and know how regarding weapons of mass destruction.

Iran's diplomatic pleas to stop arms sales to Iraq were ignored by the U.S., largely because of lobbyist influence exerted by profiteering arms producers and because of the humiliation brought by the 444-day hostage crisis that started during the Carter administration. In February of 1979 the exiled Ayatollah Khomeini returned to replace Mohammed Reza Pahlavi, Shah of Iran. Khomeini preached rabid anti-Americanism and advocated militant actions against the U.S. On November 4, 1979 Iranian militants stormed the U.S. Embassy in Tehran and took 70 prisoners into widely publicized captivity. Iran attacked tankers with oil shipments from the OPEC producers in the region to the U.S. and European countries.

Eventually, Iranian attacks on Kuwaiti tankers in the Persian Gulf and the chemical weapon attacks by Iraq brought Western European nations and the U.S. into the matter and arms supplies were finally cut to both countries. A 1988 United Nations mandated Cease-Fire Order forced the warring neighbors to stop and although both leaders refused to sign the settlement mandate, prisoners were exchanged and Iran's attacks on oil tankers ceased.

But the Iran-Iraq war was not stopped before a million young Iranian men and boys were killed along with more than 500 thousand in Iraq's young army. Large-scale use of child soldiers had taken a severe toll. And, both countries had viciously attacked civilian populations in this war to control oil. Countless children in both countries died of malnutrition and waterborne diseases. The cost to buy military supplies amounted to an estimated $500 billion for each side. Arms dealers had reaped a trillion dollars from Iraq's bloody efforts to grab Iran's oil. According to an Iraqi scientist that worked on the project before he fled from Iraq, Saddam Hussein subsequently spent more than $10 billion in a crash program to buy or develop atomic bombs that could be delivered by medium-range missiles that he acquired.

TABLE 9.1

TOP OPEC OIL RESERVES

OPEC COUNTRY	OIL RESERVES (BILLION BARRELS)
Saudi Arabia	260
Iraq[1]	110
Kuwait	95
UAE	95
Iran	92
Venezuela	66

1. Recent geological reports indicate that Iraq may have substantially larger oil and gas reserves.

TABLE 9.2

TOP NON-OPEC OIL RESERVES

OPEC COUNTRY	OIL RESERVES (BILLION BARRELS)
Russia	49
Mexico	27
China	24
United States	22
Kazahkstan	16
Norway	10

In 2002, Danish authorities brought charges against Nizar al-Khazaji the former head of Iraqi armed forces for chemical weapons attacks on Iraqi Kurds in the 1980s and for chemical weapons attacks in Iran during the 1980-1988 conflict. Khazraji, the highest-ranking officer to defect from the regime of Saddam Hussein, denied the charges and appealed the action against him.

131

In 1999 Danish authorities granted Khazraji the right to reside in Denmark because he faced the death penalty if he was forced to return to Iraq. Khazraji is charged in Iraq with sedition and treason. Khazraji, who was named as a candidate to replace Saddam Hussein, claimed that the charges against him were part of an Iraqi Secret Police plot to discredit him and disable his efforts to overthrow the Hussein regime.

Iraq invaded and occupied Kuwait in 1990. Iraq, with the largest oil holdings next to those of Saudi Arabia, had military control of Kuwait, the country with the 3rd largest oil reserves. Saddam Hussein had again used income from oil sales to finance military actions in his efforts to capture and control more oil in Iran and Kuwait. There is little doubt that terrorist leaders financed by Ayatollah Khomeini, Saddam Hussein, and Osama Bin Laden promote religious and cultural prejudices to inflame resistance to OPEC oil sales that are increasingly needed by industrialized countries.

Bin Laden's success in the recruitment and training of guerrillas and terrorists has largely been due to his ability to find additional financial backing for his violent efforts to control more than half of the world's remaining oil. Power to exploit chronically underemployed "surplus" youth by wars and terrorism comes from his ability to buy weapons from industrialized countries. Bin Laden's only significant job development programs for Afghanistan was appointments of bullies to delve out violent punishment for accused transgressions of religious directives, terrorism training, and for protection of growers and traders of narcotics.

One of the most effective strategies of terrorists is control of oil production. OPEC controls over 75% of remaining oil reserves but releases only about 40% of world production. In a few years OPEC will control over 90% of the world's remaining oil. Bin Laden and others who exploit cultural prejudices are trying to get religious militants to overthrow leaders who are favorable to Western culture in Saudi Arabia, Kuwait, Iran, Iraq and other OPEC countries.

So far, weapons of mass terrorism and destruction such as poison gases and biological agents seem to have been used by Saddam Hussein against his own people and during his war with Iran. Taliban control in Afghanistan was based upon the warlord culture of the area. However, noting the determination shown between attacks on the World Trade Center in 1993 and 2001, future acts of terrorism to inflict large-scale disruption in the U.S., Great Britain, Australia, India, Spain, Mexico and in other non-Islamic culture centers seem likely.

In the last century, six million children were seriously injured or permanently disabled by militant conflicts. Two million children have been killed in the last decade. Nine out of ten of the world's children live in developing countries where 226 million children under five years of age suffer stunting and wasting due to malnutrition, and the violent death of a child causes relatively less concern than in more developed countries.

CHILD ABUSE BY ILLICIT DRUG TRADE:

Next to waterborne diseases and malnutrition, which might be thought of as diseases of ignorance, one of the most harmful diabolical plots against children and their families is to introduce children to addictive drugs. Twenty-three countries are considered major producers of illicit drugs. In alphabetical order these countries are: Afghanistan, the Bahamas, Bolivia, Brazil, Burma, People's Republic of China, Columbia, Dominican Republic, Ecuador, Guatemala, Haiti, India, Jamaica, Laos, Mexico, Nigeria, Pakistan, Paraguay, Peru, Thailand, Venezuela, and Vietnam.

9.5

In many instances, children in these twenty-three counties are used as slave labor to grow the illicit poppies, coca, and marijuana. Children in countries with discretionary family income are targeted as customers of the narco-terrorists who supply illicit heroine, cocaine, marijuana and other drugs.

Sociopathic, mad-at-the-world personalities like Osama bin Laden and Saddam Hussein demonstrate how to be diabolical leaders. Instead of using his wealth to make a better world for so many hopeless families, Osama bin Laden's marketing plan called for development of increased heroine trade to industrialized countries and training of terror cells to blow up the gains that Civilization has made. Children in industrialized countries are the targets of drug exports from Afghanistan and other areas that host bin Laden's henchmen. Afghanistan sourced more than 70% of the world's heroine in 2000; Burma is the second largest producer. Laos is the world's third largest heroine producer.

Trafficking of illicit drugs particularly heroine and cocaine are the most profitable activities of most terrorist organizations. It is estimated to produce $400 billion in annual revenues. Drug sales support terrorist activities including drive-by shootings, car-bomb attacks, violent intimidation, and guerilla warfare. Narco-terrorists organizations include Al Qaeda, Hizballah, the Revolutionary Armed Forces of Columbia (FARC), National Liberation Army (ELN), Shining Path, The Palestinian Islamic Jihad, Islamic Movement of Uzbekistan (IMU), Kurdistan Workers' Party (PKK) Basque Fatherland and Liberty (ETA), Liberation Tigers of Tamil Eelam, and Abu Sayyaf Group. They supply drugs to peddlers who distribute to children in industrialized countries and use the income to exert their political agendas, often by terrorist activities and bribery.

To make sure life would be universally miserable for children and their parents, bin Laden's Al Qaeda gangs advocated cruel hardships for women who followed his edicts and death or other severe punishments for those daring to not comply with his oppression of personal freedom.

According to statistics compiled by the Child Welfare League of America, illicit drug use by youth is almost double the 1992 rate. Every untreated addict will cost society an estimated $43,200 annually. Every year about 500,000 U.S. babies have been abused by having prenatal intake of illicit drugs that cause impaired motor skills, delayed language development, hyperactivity and related behavioral problems. Continuation of current trends in the U.S. will cause child-welfare and substance abuse costs to be more than one trillion dollars over the next 20 years.

Saddam Hussein hosted and shielded narco-terrorism leaders. Except for his hatred of America, it is hard to understand why Hussein, with more than 3.9 trillion dollars worth of oil to market would help narco-terrorists but he seemed to do so. In spite of ruling over enormous potential oil wealth, Hussein seemed intent on doing everything he could for decades to alienate oil customers. He disaffected potential customers within Iraq by wielding punishment for their existence. Hussein attacked Iran and Kuwait, his fellow members in OPEC. Saddam's vanity, insatiable greed, and lust for power became legendary.

Instead of buying expensive weapons, Saddam Hussein's regime could have wisely used a small fraction of Iraq's 3.9 trillion dollar oil wealth to hire the under- employed of the Middle East to build and install Renewable Resources Parks throughout Iraq and neighboring countries. By now, the cradle of Civilization could have become the cradle of the Solar Hydrogen Civilization. Saddam Hussein would have gained the respect and admiration that Civilization bestows upon truly great leaders.

However, Saddam Hussein's continuing treachery apparently made it seem necessary to sacrifice giant oil-production concessions to Russia, China, France and Germany to gain their opposition to U.N. resolutions to halt his military weapons purchases and forcibly disarm Iraq's military threat to neighboring countries. Instead of using oil income to build infrastructure and support exemplary renewable economic development, Hussein used Iraq's enormous oil wealth to cause decades of hardship and trouble for his country, his neighbors, and the world.

RESPONSIBILITY FOR OVERCOMING CHILD ABUSE AND TERRORISM:

9.6 How will peace-loving Moslems, Christians, Buddhists, Hindus, and those of other faiths and beliefs overcome child abuse and terrorism? Military actions against terrorist training camps have helped to keep them off balance. It is equally important to remove these camps and their leaders to prevent them from being a continuing influence on the attitudes and characters of developing children. Tighter enforcement of money laundering that supports terrorists will reduce their capabilities. Prevention of child abuse requires much more education of would-be parents and improvement of job opportunities.

About 8,000 child deaths per day due to six dreaded child-killing diseases (measles, tetanus, whooping cough, tuberculosis, polio, and diphtheria) are now prevented by humanitarian actions that provide widespread vaccinations. These 3 million lives that are saved each year desperately need good food to avoid the perils of malnutrition including stunted physical and mental development. They need wholesome environments and high-quality education.

Upon maturing they need good jobs. If these needs are not met, some 8,000 children per day will be subject to the persistent syndrome of child abuse by neglect and disdain that faces the larger portion of the world's growing population of children. Acceptance of further child abuse and conscription into terrorist support will grow if these needs are not met with opportunities to participate in the worldwide achievement of sustainable prosperity without pollution.

At the time the global population passed six billion, one billion were teenagers under 18. This emerging job force including the 300,000 teen-children exploited in military efforts are needed in much more healthful and beneficial pursuits that ultimately improve the environment for virtually every form of life on Earth. They, along with the billions of other youth are needed in schools and training programs to supply job-ready persons to accomplish Civilization's Grand Purpose of Sustainable Prosperity.

The global economy needs virtually full employment in every country to overcome dependence upon burning over one million years of fossil accumulations each year. Contrary to prevailing theories by economists and advice of business investment analysts, rising retail prices for gasoline do not signal a long-term opportunity for investments in fossil fuels.

In order to achieve long-term economic benefits it is essential to direct investment to education and technology transfers for production of renewable energy and materials production instead of mining away Earth's remaining fossil reserves. This worthy investment purpose must include prioritized development of practical solutions to the problems indicated in Tables 9.3 and 9.4.

It is the responsibility of peace-loving citizens in every country to refuse to support terrorists and those who encourage terrorism. Peace-loving mentors of every respectable trade and profession must identify and advocate the benefits of sustainable, clean, safe, and practical energy-conversion and materials-development technologies. This advocacy must be matched with hard work and dedication as needed to provide job opportunities for everyone who seeks to be part of the revolution that achieves better outcomes than continued strife and pollution caused by diminishing supplies of fossil reserves.

New ventures that make, use, and sell something that advances renewable resources need to be formulated by teams that combine experienced engineers, accountants and other professionals with newly educated persons. These new businesses must successfully compete for capital investment and market support in comparison with those who depend directly or indirectly upon fossil resources.

What must happen next in Iraq, Iran, North Korea, and other troubled lands is for concerned parents to set aside hatreds, prejudices, ego-power maniacs and greed-driven dictators so they

Preventing Protracted Child Abuse and Self Destruction

Saving children for a meaningless life of hardship amounts to cruel and protracted child abuse. The 8,000 children that are saved each day by vaccinations against dreaded diseases will need worthy jobs to earn satisfaction that their lives are essential and esteemed in Civilization's accomplishment of the Solar Hydrogen Civilization. It is imperative that far larger, more distributed, and sustainable supplies of solar, wind, wave, falling water, and biomass resources replace present dependency on diminishing supplies of fossil resources. Every person on Earth has a job to accomplish sustainable prosperity and to do less is a form of self-destruction.

TABLE 9.3

INVESTMENT, EDUCATION AND JOBS NEEDED TO OVERCOME PROBLEMS

1. More than a trillion dollars have been spent for military efforts in Iran, Iraq, Kuwait, and Saudi Arabia on oil wars since 1980. This amount of capital could have launched new ventures that would have hired most of the Middle East's unemployed to build pipelines, Renewable Resources Parks, new communities, polymer plants, industrial parks and much more. This model of peaceful progress could have been provided instead of continuing terrorism and conflict throughout these countries and Africa.

2. Job development requires $250 billion for financing new ventures that will make, use, and/or distribute renewable energy, clean water, and materials. It is time for Iraq to become the Solar Hydrogen epicenter for sustainable economic development throughout the Middle East and Africa and overcome much of the conflict, growing poverty, illness, and misery summarized in Tables 9.3 and 9.4.

3. Providing sustainable economic development is the most important Nation Building Task and it is essential for overcoming rapidly growing problems in the Middle East, Africa. Asia and South America

SOCIAL DILEMMA	NUMBER EFFECTED	SOLUTION INDICATED[1]
Persons Without Clean Drinking Water	1.1 Billion and Growing	Renewable Resources Education and Job Development
Persons Without Basic Sanitation	2.8 Billion and Growing	Renewable Resources Education and Job Development
Daily Deaths Due to Waterborne Diseases	25,000 (Largest Cause of Sickness and Death)	Renewable Resources Education and Job Development
Persons Living On Less Than $1.00US/Day	1.3 Billion In Poverty And Growing	Renewable Resources Education and Job Development
Women and Children In Extreme Poverty Due to Wars, Violence, Illness, and Natural Disasters	48 Million and Growing (75% of the World's Poorest Population)	Renewable Resources Education and Job Development
Child Deaths (Under Age 5) Due To Malnutrition	6.6 Million	Renewable Resources Education and Farm Job Development
Child Stunting and Wasting (Under Age 5) Due To Malnutrition	226 Million In Developing Countries	Renewable Resources Development and Farm Jobs
School-Aged Children Not Receiving Education	130 Million	Basic Education Required
Illiterate Persons In Developing Countries	850 Million	Job Opportunities For Literate Persons
Illiterate In Developing Countries Who Are Undernourished	840 Million	Encourage Reading To Grow Foods and Eat Better

PROJECTED POPULATION CHANGES BY 2050

TABLE 9.4

EXPECTED WORLD POPULATION CHANGES

ASIA	Growth from	2.5 - 5.4	BILLION
AFRICA	Growth from	.80 - 2.0	BILLION
LATIN AMERICA	Growth from	.49 - 0.8	BILLION
NORTH AMERICA	Growth from	.41 - .54	BILLION
EUROPE	Decline from	.73 - 0.64	BILLION

can devote their energies and the renewable resources of their countries to accomplishment of locally appropriate versions of The Solar Hydrogen Civilization. It is time to overcome child abuse that is promulgated by greed, hateful prejudices, fraudulent con-men, and ego-power maniacs.

SUSTAINABLE FARMERS, MANUFACTURERS, AND HOMEMAKERS WILL BE ESSENTIAL TO OVERCOME CHILD ABUSE AND TERRORISM:

Farmers of the Solar Hydrogen Civilization in every land will become important energy producers by converting solar, wind, falling water, and biomass resources into hydrogen, methane, and electricity. In addition to being the essential providers of food, farmers and their philosophy of peaceful agriculture will gain importance as renewable energy supplies are increasingly produced by area farms. Farmers will use hydrogen to power tractors and other farm equipment to become more cost-effective producers. Hydrogen will enable existing engines to produce more power and last longer. By using hydrogen to eliminate small particulates from petrol-burning engines that now cause lung and heart diseases, farmers will maintain higher productivity and enjoy a better and longer life.

9.7

Conversion of sequestered carbon into filter media, refractories, products reinforced by carbon fibers and many other products will provide employment by industries that enter into contracts to receive pipeline supplies of methane from farmers. Pipelines will be built to connect farm areas to cities on every continent to provide efficient storage and delivery of ethane, methane, and hydrogen.

Renewable methane will be dissociated into carbon to make durable goods and hydrogen. Manufacturers will become sustainable by entering into electronic contracts to receive renewable electricity and methane from farmers and other entrepreneurs. Methane will be split to produce hydrogen and carbon to manufacture the air and water filters, solar collectors, wind turbines, falling water generators, and biomass digesters that are needed in great quantities around the world. Manufacturers will provide high-efficiency total energy systems to businesses and homemakers.

New products that are needed to overcome the dilemma of dependence upon fossil fuels include annual assembly-line production of at least 10 million wind generators, 10 million solar dish energy converters, 10 million wave machines, 10 million anaerobic waste converters, 10 million home-size reversible fuel cells, 10 million TESI appliances, and hydrogen conversion kits for at least 10 million vehicles. Each of these industries represents economic development equivalent to the U.S. automotive industry.

Homemakers will provide sustainable sanctuaries for family members. Pollution-free air will be provided by hydrogen engines that clean the air. Indoor air, once characterized by fumes from paint, carpeting, and furniture will be cleaner and more healthful because of new material selections that have virtually no out-gassing. TESI units will clean air to produce opportunities for much better health and peace of mind.

Home study opportunities that provide educational information by web courses will enable faster learning and quick access to a wide array of information resources. Home study must be made practical by TV and Internet programs for families in less developed regions. More than 90% of the global population growth will continue to occur in less-developed regions. The fastest way to develop awareness of job opportunities for the emerging job seekers shown in Tables 9.3 and 9.4 will be through cartoon booklets, TV and internet educational programs, and in home-study courses.

SKILLED LABOR WILL ACCOMPLISH SUSTAINABLE PROSPERITY:

9.8 In order to produce, distribute, and utilize renewable hydrogen and electricity to replace the fossil equivalent of 200 million barrels of oil that are burned daily, the global economy will require many new jobs. In addition, the developing countries need at least this much more renewable energy to meet essential needs for adequate living standards. Gainful participation by virtually all of the persons now alive on Earth in the Renewable Resources Revolution is needed. An average of one out of every six persons will be producing, distributing, selling, servicing, and in other ways be involved in the sustainable businesses that facilitate the coming peaceful renewable energy and materials revolution. Five of the six persons will utilize renewable energy and materials in more desirable ways than could be accomplished by the use of fossil resources.

Integration of the efforts of all six persons in the philosophy of investing in and working to achieve sustainable prosperity will overcome the syndrome of shortsighted globalization that exploits labor, squanders finite resources, and despoils the environment. Broadly described, the two main resources that Civilization depends upon, namely Nature and Skilled Labor have been mismanaged and exploited in ways that produce adverse consequences.

Economic growth has been temporarily achieved at the expense of nature as illustrated by the adverse changes due to depletion of fossil and mineral reserves, depletion of stratospheric ozone levels, greenhouse gas accumulation, soil erosion, depletion of glacial-age aquifers, water contamination by MTBE and countless other wasted substances, development of prions in the food chain, and loss of fisheries due to pollution. Such degradation of nature causes deterioration in the quality of life and undermines the satisfaction needed to justify the large expenditures of resources that have been depleted.

Skilled labor has also been persistently discouraged in the last half of the Industrial Revolution. First, skilled labor (that is the product of generations of skill building) invents and produces the inexpensive wonders that we find so desirable. Then skilled labor is replaced by lower-cost labor from populations with persistent unemployment. And as this is happening, a few of the very most skilled are asked to automate the machinery involved to virtually eliminate labor content. A competition develops between products from super-efficient, automated machinery and imported products made by low-cost labor.

But the populations with such unemployment and low-cost labor probably are not early beneficiaries of the desirable goods that they produce. Populations with the automated machinery incur dissatisfactions also. This is because these highly desirable goods from cheap-labor areas are exported to the skilled-labor population that increasingly pays for the imported goods from unemployment checks. Owners of the automated production factories are taxed more and more to pay the costs of unemployment and everyone involved suffers a loss of satisfaction in this version of the globalization process.

In order to overcome this too familiar syndrome of nature and skilled labor exploitation the following philosophical adjustments for producers and buyers are needed:

Skilled labor in developed countries must rapidly invent, develop and efficiently manufacture sufficient tools and equipment for export to developing countries to support jobs for increasing farm productivity with renewable resources including hydrogen, electricity, methane, and carbon.

Virtually every durable good must be designed for ultimate recycling after it becomes obsolete by converting its material contents into new and better products of the future through the use of renewable energy. Skilled labor of the future needs materials that produce and distribute renewable energy and that are produced by presently available skilled labor.

Toys, popular reading, and entertainment programs need to indicate how environmentally desirable products can be selected and enjoyed more than products that cause environmental degradation and loss of jobs. Such suggestions to children will enable much better career planning and will encourage better purchasing power to be developed in future markets.

Products must be designed and manufactured for extended life by refurbishing and repair when needed. Nearly every throwaway item can be designed for repairable service and doing so will develop many more worthy jobs for skilled labor.

Products with reduced energy and material contents can be designed and manufactured from emerging carbon-characterized materials. Transporting products great distances will be less attractive than locally producing products needed for local markets. Reductions in product weight generally provide improved performance and reduced requirements for operational energy. Greatly reduced life-cycle energy requirements result.

Buyers need a life-cycle index of all externalities associated with the production, repair, recycling, emissions, hazards, etc., of products. This will provide the buyer with much greater information and justification for initially paying more for products that will provide greater ultimate satisfaction.

MORE THAN TWO BILLION JOBS FOR THE GLOBAL ECONOMY:

9.9 Good job opportunities will be created in each of 6,500 Self-Help Districts to produce sustainable economies. Each Self-Help District will have a population of approximately one million persons. This represents the production density and magnitude typically needed for optimizing collection and transport of organic wastes for efficient sequestration of carbon to produce durable goods and for production of hydrogen for transportation, electricity generation, and process chemistry applications. Self-Help Districts will achieve sustainable economies by converting sequestered carbon into components, equipment, and systems that harness locally available solar, wind, falling water, waves, and/or geothermal resources into hydrogen and electricity. Table 9.5 lists typical job opportunities that will develop in many of these Self-Help Districts. Other Districts will become specialized as needed to optimize local conditions and resources.

In comparison with being recruited into a terrorist's life of discontent, anger, and violent disruption; new job opportunities that are far more exciting, interesting, positive, rewarding, satisfying and that support a long life of praiseworthy advancement are listed in Table 9.5.

TABLE 9.5

NEW JOB OPPORTUNITIES

JOB OPPORTUNITY	EMPLOYMENT PERCENTAGE
SERVICES	20
RETAIL SALES	16
ENERGY PRODUCTION	16
CONSTRUCTION	15
MANUFACTURING	13
FOOD PRODUCTION	8
TRANSPORTATION	6
WHOLESALE DIST.	4
FINANCE/INSURANCE	2

JOB DEVELOPMENT DESCRIPTIONS:

9.10 Manufacture of modular waste converters to produce methane, ethane, hydrogen and carbon from sewage, garbage, agricultural and forest wastes will employ skilled technicians. Job opportunities will be created to manufacture, market, distribute, install and service these modular units.

Production of new cash crops of methane, hydrogen, and electricity on every farm will require skilled technicians. New jobs will be created to install, service, and operate biomass waste conversion systems throughout the world.

Technicians will manufacture polymers, carbon and glass reinforcement materials, along with steel to make pipe, fittings, and valves needed to transport renewable methane and hydrogen from farms and Renewable Resources Parks to cities. Jobs will be created to make the tooling and produce these essential components along with sales and service opportunities.

Technicians will manufacture electronic contract meters for wheeling renewable electricity, methane and hydrogen from producers to customers. Marketing, installation and service personnel along with technicians in the manufacturing sector to produce the semiconductors, sensors, instrumentation, hardware and software for facilitating sales of renewable resources will provide new jobs.

New jobs will be created to manufacture, market, install, operate, and service 10 million new solar dish gensets per year. Installations in Renewable Resources Parks, on farms, and businesses create new employment to provide renewable hydrogen and electricity from solar-rich areas.

Technicians are needed to produce, sell, install, operate, and service 10 million new wind generators per year. Wind-rich areas of the world will produce renewable energy and provide expanding employment opportunities.

Similar skills are needed to manufacture, sell, install, operate and service 10 million new wave generators per year.

Manufacturing technicians, mechanics and technical sales personnel are needed to produce, install, operate and service 10 million new Total Energy System Innovation (TESI) units per year. This will create new jobs and greatly improve energy-utilization efficiency.

Technicians are needed to produce; sell, install, and service 10 million hydrogen retrofit kits per year. Installation kits for the global transportation fleet and TESI installations will produce market pull for hydrogen from every farm and Renewable Resources Park.

Technicians are needed to manufacture, install, operate, and maintain 10 million reversible fuel cells per year.

MARKETING PRODUCTS THAT ENABLE SELF-HELP DISTRICTS ACHIEVE SUSTAINABLE PROSPERITY:

In addition to the technical jobs that will be generated to produce, transport, store, and utilize solar hydrogen; friendly, people-oriented marketing specialists are needed to favorably present and sell renewable energy products to the public. Marketing personnel in each of the 6,500 Self-Help Districts of sustainable progress will have a mission, plan and purpose to provide products that facilitate accomplishment of sustainable prosperity.

9.11

In each self-help district, products that will be marketed will be locally appropriate and designed to produce the highest socioeconomic results. Sales personnel will be marketing what it takes to provide a much higher standard of living for the young and restless, middle aged, old and wise, and virtually all other descriptions of humanity. They will market what is needed to overcome widespread child abuse due to malnutrition, disease, apathy, wars, terrorism, and drug abuse.

THE SOLAR HYDROGEN ECONOMY

Hydrogen stars shine energy throughout the universe, the relatively small amount of radiation that reaches Earth allows hydrogen to power every living cell. The emerging Solar Hydrogen Economy will save Civilization from looming economic collapse. And, the Solar Hydrogen Economy will progressively reverse many of the environmental hardships caused by the fossil fuel age.

What is it? The Solar Hydrogen Economy will utilize energy from the sun along with derivatives of solar energy such as wind, wave, falling water, and biomass resources to produce goods and services. The fuel of the Solar Hydrogen Economy is hydrogen that is produced from water and/or other hydrogenous compounds by microbes, thermal dissociation and/or electrolysis.

How can I be a part of it? Photovoltaic panels, wind generators, falling-water generators, and other devices that harness renewable energy are usually sized to meet peak electrical loads of a dwelling or a community. Sizing systems to meet peak demand means that there will be a great deal of electrical generation capacity in surplus between peak load requirements.

CHAPTER 10.0

Electrolysis of water to produce pressurized supplies of hydrogen and oxygen offers a practical way to utilize the surplus generating capacity and enter the Solar Hydrogen Economy.

When can I have it? Pressurized gas, cryogenic liquid, and ambient temperature hydride storage systems are well proven hydrogen storage technologies. It is practical for many to operate ordinary engines on hydrogen. Hydrogen fuel cells are beginning to power homes, vehicles and appliances.

LESSON LEARNED FROM PLAYING MARBLES: *You have to play in the right game to win.*

Before entering school I tried to keep up with my three older siblings and their friends. One of the few games that they would let me join was marbles. Why not, I was a good loser who kept turning earnings into new supplies of marbles.

When I started to school my skills at buying and playing marbles were more developed than most first graders, and soon parents were complaining that something had to be done about the boy who was taking home too many marbles. I was instructed to only play with the second and third graders, or I would not be allowed to take home any more marbles. If complaints came from the parents of second or third graders, I would have to get the fourth or fifth graders to let me play with them. My teacher said, *"You have to play in the right game or you cannot take the marbles that you win home with you."*

This lesson in civics has been reconsidered through the years. Was it harmful to learn the ropes from the big kids that let me play marbles with them? Was it any better to protect my fellow first graders from losing their marbles? Did the kids who had their parents and the teacher intervene learn more about civics and skill building than I had learned?

DEPENDENCE UPON MASS PRODUCTION:

Mass manufacturing is necessary to provide enough food, clothing, and housing along with an encyclopedia of consumer goods that are too often taken for granted. Farmers eat factory prepared cereals poured from appealing packages instead of spending time to grind the grain, measure out the recipe, bake and sugar-coat the flakes. Eating the mass-produced store-bought concoction saves time so each farmer can efficiently produce enough food commodities to feed 60 persons in the cities.

10.1

In the cities cars that are mass-produced quickly travel from place to place. Only a very small percent of Earth's population is employed to efficiently manufacture 40 million new cars each year. If each of the 40 million persons tried to make his own car it would take decades to produce a small fraction of the cars that are needed each year, and the cost per car would be astronomical as would the cost of insurance for these special designs and uncertain quality controls.

Similarly if carpenters had to saw every timber, split the shingles, chisel every stone, make every screw, and cut each nail, construction would be very slow and enormously expensive. Efficient mills cut timbers to size, bricks are mass-produced in automated plants, streams of screws and nails spew out of automatic forming and heading machines at 60 MPH. Homes can be rapidly built and cost much less than if they were assembled from hand-made components.

Efficient mass production of solutions to the dilemma of fossil dependence is necessary because the present population is so large. We must organize mass production of solutions like our lives depend upon it... they do. Civilization depends upon it. Some of the facilitating inventions that must be mass manufactured by new ventures are vehicle conversion kits, compact hydrogen storage tanks, modular waste to hydrogen and carbon units and electronic-contract meters.

Solar Hydrogen is the right game but what rules must we learn to win the game to provide Sustainable Prosperity in every community of the world?

We must organize mass production of facilitating technologies like our lives depend upon it... in many ways, our lives do depend on it... and Civilization depends upon it.

THE FORT COLLINS DECLARATION:

Fort Collins had its first fall snow on September 23, 2000. It was beautiful to see the falling snow from the hotel window. Cars parked in front of the hotel slowly covered with snow, trees turned white, and streetlights twinkled through the falling flakes that gathered to transform the landscape from earth tones to white.

10.2

Maurey Albertson had arranged a convention to bring the grassroots hydrogen people from Hydrogen Now, the American Hydrogen Association, and the International Association for Hydrogen Energy together to plan a much larger meeting for the next year. Nicko Kangelaris joined me in promoting the notion that advancement of Solar Hydrogen was too important to leave until next year. We set out to write the Fort Collins Declaration to unite people of every continent in the quest for sustainable prosperity.

We passed out preliminary drafts and requested suggestions. Soon there were many others interested in helping with the grammar and proposing ways to make the Declaration known to the world. On September 24, 2000, The Fort Collins Declaration was brought before the convention for reading and signing. Nicko Kangelaris III, Roy McAlister, Dr. T. Nejat Veziroglu, and Dr. Maurey Albertson were the first to sign. After the signing ceremony, it was decided that the following year at this same date, this group would meet with the National Hydrogen Association, NREL (National Renewable Energy Laboratory) personnel and numerous others for the purpose of advancing the goals of the Declaration. The convention adjourned with determination to use the Declaration as the theme of the 2001 convention.

After the September 11, 2001 terrorist attacks on the World Trade Center Towers in New York City, the Pentagon attack, and downing of another commercial airliner, all planes were grounded. The 2001 convention had to be cancelled. Was that to be the end of the Declaration?

World War III threatens as an extension of the first two world wars because it is shaped by continuing hostile intentions to control oil and other critical resources. Desert Storm paused with an agreement to control oil sales from Iraq but it continued with air strikes to enforce the intentions of U.S. and other countries to prevent Iraq from taking over neighboring oil fields by promulgating dogmatism, terrorism, and mass destruction. But even after Saddam Hussein and Osama bin Laden disappeared, terrorists and rioters continued to disrupt peace.

The Fort Collins Declaration is a call to all mankind to help bring about the Solar Hydrogen Economy as a much better solution to pressing needs for energy sufficiency and pollution avoidance. It is a much better solution than continuing with military intervention over diminishing fossil resources.

Solar Hydrogen *is the right game…every player can win.* Everyone who breathes wins. Workers have much better jobs and the renewable energy and energy-intensive products and services that we will produce are anti-inflationary. Capitalists will have better investment opportunities. Economists can be more important by writing books on wealth-expansion economics. Civilization's need for sustainable wealth expansion has been ignored as we muddled in the temporary economics that we have based on fossil fuel depletion.

When September 23, 2002 came along I remembered the night that we wrote the Fort Collins Declaration. The new Millennium gave us hope to start the world on a new quest for sustainable prosperity. I am as dedicated, more inspired, and know that we can achieve the Solar Hydrogen Civilization.

Make your pledge today to support transition to the Solar Hydrogen Civilization. Find your place in the "Call to Action"… and quickly learn how to solve our most urgent problems of energy sufficiency and pollution.

THE FORT COLLINS DECLARATION
September 24, 2000

The undersigned citizens of the Earth are resolved that civilization must find peaceful ways to overcome the problems of economic demise and environmental pollution attributable to dependence upon the growing annual combustion of fossil reserves that took millions of years to accumulate. With global consumption outpacing nature's production exponentially, following our current path equates to economic suicide and runs the grave risk of making our planet uninhabitable by present and future human populations.

10.3

We realize that the Industrial Revolution has been predicated upon exponentially increasing expenditures of the earth's resources including fossil reserves, potable water, and other natural capital of the only environment known to exist in the universe that can support Civilization's present and future human populations. One result is that the Industrial Revolution has provided tremendous technological advances that have facilitated the exponential increase of human population from one billion to six billion persons since the beginning of the commercial exploitation of fossil reserves as fuels.

The Industrial Revolution has also improved our standard of living. Coupled with the high velocity of technological advances, the Industrial Revolution continues to increase our standard of living at an even higher rate. As our standard of living improves by present practices, stresses on earth's natural capital increase. Taking the current pace of worldwide technological and industrial advancement into consideration, earth's fossil reserves are being exhausted faster and faster.

We maintain that it is beneath human dignity for any government to risk the lives of its citizens serving in the armed forces to protect transportation bottlenecks of these inefficient fossil reserves situated in foreign, and often hostile, lands. It is also irresponsible of any government to risk the lives of innocent bystanders who are guilty of nothing but their geographic location during unnecessary disputes, crises, and international incidents revolving around said bottlenecks.

We also realize that "Solar Hydrogen" (meaning hydrogen produced by harnessing solar energy and/or its derivatives including falling water, wind, wave, and biomass resources) offers a practical way to continue the progress of the Industrial Revolution, improve the standard of living for all living things on Earth, and promote global prosperity without pollution and conflict.

We realize that virtually any of the world's 800 million existing engines can be rapidly converted to operation on hydrogen with the benefit of actually cleaning air that passes through the engine. We want an increased return on the enormous investment that has been made to mine, refine, and manufacture the world's existing engines by using hydrogen to extend their lives and increase their performance. Fueling existing engines with Solar

Hydrogen will create the infrastructure needed to facilitate advancements of fuel cells, advanced technology engines, and other innovative applications.

We further realize that it is essential to utilize petrocarbon reserves to produce durable goods instead of incurring the economic losses caused by burning them. One gallon of average crude oil can be used to produce over $35.00 worth of efficiently recyclable products such as polymer-based components for improving the performance and durability of such things as equipment for farming, mining, and manufacturing; computers, vehicles, roads, clothing, and homes.

THEREFORE, THE UNDERSIGNED ARE COMMITTED TO THE DEVELOPMENT AND IMPLEMENTATION OF HYDROGEN SYSTEMS, GOODS AND SERVICES TO FACILITATE A SUSTAINABLE GLOBAL ECONOMY WITHOUT POLLUTION.

CALL FOR ACTION:

1. *Legislators and Policy-makers* must provide the focus and leadership needed by Earth's human population to achieve a sustainable economy by repealing subsidies that hinder advancement of renewable resources and enact laws and regulations to encourage and support the establishment of the Solar Hydrogen Energy System.

2. *Researchers and Developers* must provide scientifically proven options for continuously improving production, storage, delivery, and safety concerning widespread utilization of Solar Hydrogen.

3. *Scientists and Engineers* must use the best scientific methods and practices to discover, design, and mass produce what is needed for rapid transition to the Solar Hydrogen Economy.

4. *Entrepreneurs* must use the best business development practices to facilitate rapid transition to the Solar Hydrogen Economy.

5. *Educators* must provide students with the information needed to enter the Solar Hydrogen job market with the essential skills, dedication, and attitude required to elevate all communities of the world to sustainable prosperity without pollution.

6. *Parents* must provide their children with the awareness that human potential depends upon personal responsibility to protect the environment and to provide a high standard of living based on sustainable resources.

7. *Economists* must include **natural capital** and **opportunity costs** as essential aspects of economic models, wisdom, and decision-making. Illustratively, it is an unacceptable opportunity cost to burn one gallon of oil compared to producing $35.00 worth of recyclable durable goods. Even if the sale price of a gallon of oil escalates to $10.00 per gallon and the price of polymer compounds for producing improved computers, automobiles, and housing remains at $35.00 per gallon, the unacceptable opportunity cost of burning a gallon of oil will be $25.00 per gallon.

8. *Investors, Businesses, Farmers, and the like* must develop the expectation and demand for business opportunities that add wealth to the inventory of goods and services that are available to the Citizens of Earth. Much of the progress in this regard can be expected from investment in energy-intensive production of goods in which the energy and indeed the material constituents are derived from solar, wind, wave, falling water, and biomass resources.

9. *Communication experts* must help develop widespread public knowledge about Solar Hydrogen and inspire action to achieve sustainable global prosperity.

SOLAR HYDROGEN ECONOMIC DEVELOPMENT CORRIDORS:

What is needed for significant markets to be served is achievement of the economy of scale that distinguishes the Industrial Revolution from every other milestone of human progress. Practical evolution of the present stage of the Industrial Revolution to the Solar Hydrogen Economy will be expedited by establishment of corridors where Solar Hydrogen is made available for essential activities including farming, electricity production and refueling vehicles that are converted to operation on hydrogen.

10.5

Virtually any roadway can become a travel artery of a Solar Hydrogen Corridor. The roadway could be your nearest highway or Highway 1-95, I-40, I-5, or Route 66. What will make it work the soonest is distribution of renewable methane through existing natural gas pipelines to industrial parks that are located along the defining roadway. These industrial parks will become Renewable Industries Parks where increasing percentages of renewable methane that is delivered by natural gas pipelines will be converted into electricity, hydrogen and carbon products.

The hydrogen will be sold at convenience markets to motorists who will refill canisters that fuel barbecues, motorboats, contractor's engine-generators, and many other applications. Carbon products that are produced in the industrial parks will include fibers that are stronger than steel and lighter than aluminum, diamond plating, and activated-carbon filter media for improving the quality of water and air.

Methane for distribution through the existing natural gas pipelines will be produced from virtually anything that is now burned or allowed to rot away in landfills and on farms or in forests. Sewage, garbage, crop wastes, and manure will be converted into methane for distribution through natural gas pipelines to industrial parks along the Solar Hydrogen Corridor.

Renewable methane can be mixed in all proportions with natural gas without requiring any change in the natural gas distribution system.

Each year more carbon enters the atmosphere from decay or burning of biomass to form carbon dioxide and/or methane than from all the coal that is mined and burned. Forest fires, swamp gas, farm wastes, rotting garbage, sewage disposal operations and other events that cause biomass to burn or decay produce

Industrialized communities have a moral responsibility to invest what is required to expedite automated mass production, distribution, and installation of inventions for enabling every community of the world to harness solar, wave, wind, falling water, and/or biomass waste resources to create sustainable prosperity.

enormous amounts of methane and/or carbon dioxide. Biomass resources present opportunities for production of renewable carbon, hydrogen, and soil nutrients.

Figure 10.1 shows the Carbon Cycle of the Industrial Revolution. Humans have changed the amount of carbon that is available for the carbon cycle by adding carbon from fossil deposits that were buried for more than 60 million years. We now burn the fossil equivalent of about 9 billion gallons of oil each day and this has ramped up the carbon dioxide content of the atmosphere. Presently there is about 30% more carbon dioxide in Earth's atmosphere than at any time in polar snow core records that date back some 160,000 years.

Sequestering carbon from biomass to produce durable goods will reverse the build-up of carbon dioxide in the atmosphere. At Renewable Industries Parks, methane from renewable sources will not be distinguishable from methane from fossil sources. But the practice of separating hydrogen from carbon and using the carbon to produce durable goods will play an increasingly important role in saving Civilization from the embarrassment of wasting the old and newly photosynthesized carbon and suffering the atmospheric changes that harm us.

When natural gas and other fossil fuels are burned in conventional furnaces and engines, the carbon content becomes carbon dioxide and is discarded into the atmosphere. A gallon of liquid fuel such as oil or gasoline converts about 5.4 pounds of carbon into about 19.8 pounds of carbon dioxide. About 122 standard cubic feet of natural gas can deliver as much energy as a gallon of gasoline. Burning this much natural gas will convert about 3.83 pounds of carbon into 14 pounds of carbon dioxide.

FIGURE 10.1

The Carbon Cycle of the Industrial Revolution: Carbon dioxide including carbon from fossil burning is taken by plants to produce plant tissues, the plant tissues burn or decay and release the carbon dioxide.

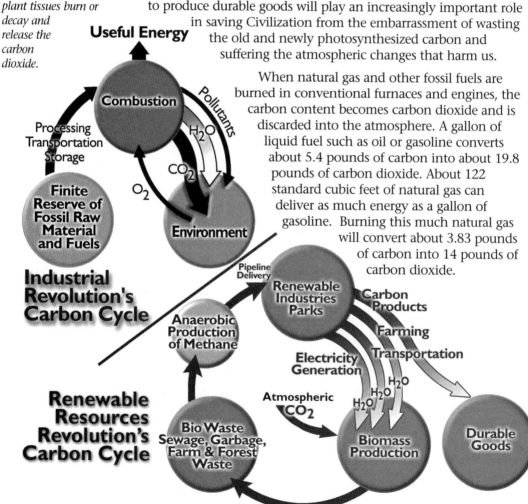

Useful Energy

Combustion

Processing Transportation Storage

Pollutants

H_2O

CO_2

O_2

Finite Reserve of Fossil Raw Material and Fuels

Environment

Industrial Revolution's Carbon Cycle

Pipeline Delivery

Renewable Industries Parks

Carbon Products

Anaerobic Production of Methane

Farming

Electricity Generation

Transportation

H_2O

H_2O

H_2O

Atmospheric CO_2

Renewable Resources Revolution's Carbon Cycle

Bio Waste Sewage, Garbage, Farm & Forest Waste

Biomass Production

Durable Goods

Each pound of burned carbon produces about 3.67 pounds of carbon dioxide. Burning hydrocarbon fuels incurs the opportunity cost equivalent to producing 3 to 5 pounds of durable goods. We must avoid adding more carbon dioxide in an already polluted atmosphere along with the climate changes that it causes. Consider the economic advantages of using the carbon to make 3 pounds of carbon-reinforced transportation components that are stronger, safer, lighter, and that resist corrosion better than aluminum, titanium or steel. Lighter and safer vehicles will have far better fuel economy.

Producing durable goods from carbon extracted from biomass and fossil sources will increase profits and produce far more jobs than continuing to allow such carbon to enter the atmosphere as carbon dioxide, carbon monoxide, methane or some other pollutant. Solar Hydrogen Economic Development Corridors will quickly produce more employment, superior products, increased consumer goodwill and much more investor confidence. Hydrogen will be available to improve the efficiency, power production, and the life of existing engines. Development of fueling centers and home production of hydrogen to fuel existing engines from renewable resources will create the Solar Hydrogen infrastructure required to support fuel cells as they arrive.

REALLY CLEAN COAL MAKES DURABLE GOODS:

One way to actually make coal live up to the promise of "clean coal" is to make it into durable goods instead of burning it. Methane can be made from coal by reacting hydrogen from renewable resources with the carbon in the coal. This will allow the carbon that has been thus transported from coal mines to be converted into durable goods at the Renewable Industries Parks that supply the hydrogen to meet local fuel needs. Equations 10.1 and 10.2 show the overall processes:

$$Coal + Hydrogen \rightarrow CH_4 \qquad \text{EQUATION 10.1}$$

$$CH_4 + HEAT \rightarrow C + 2H_2 \qquad \text{EQUATION 10.2}$$

Process chemistry for producing methane from hydrogen and coal facilitates clean-up of the coal by removal of sulfur, ash and other impurities. Clean methane that is produced can be transported to industrial parks where it will be made into carbon products and hydrogen as summarized in Equation 10.2. Sulfur removal and purification can be accomplished by using the hydrogen to form gaseous hydrogen sulfide which is then separated from the solid residue remaining. Hydrogen sulfide is then decomposed into hydrogen and purified sulfur. Equations 10.3 and 10.4 summarize the process:

$$Coal + Hydrogen \rightarrow H_2S \qquad \text{EQUATION 10.3}$$

$$H_2S \rightarrow S + H_2 \qquad \text{EQUATION 10.4}$$

Ash can readily be removed at the stage that coal is decarbonized and sulfur is removed. This provides an overall process regime in which carbon from coal is utilized for much higher value purposes than as a fuel. It facilitates delivery of hydrogen and clean carbon to Renewable Industries Parks along Solar Hydrogen Corridors.

Solar Hydrogen Corridors will be preferred places to work and live because the jobs are better and the environment is improved. Greater job satisfaction and improved mental well being will be produced by awareness that human activities are improving the chances for good health and sustainable economic development.

Carbon products including doped diamond semiconductors will revolutionize microelectronics by operating faster, facilitating increased operations per second, and remaining stable at higher operating temperatures. In addition, carbon crystals provide better heat transfer and optical characteristics to provide new horizons for combining optical and electronic functions.

Newly emerging super-hard and slick carbon coatings will provide energy savings by reducing friction and heat generation between relative motion components. This advantage can be utilized in industrial equipment, farm implements, and in transportation engines, transmissions, and differentials.

Thermodynamic analysis shows that carbon production from bio-wastes and hydrocarbons is more attractive than burning the carbon and then investing much more energy to sequester the carbon dioxide that is formed. Removal of carbon dioxide from the exhaust stacks of a coal burning power plant has been extensively studied. It is estimated that removing carbon dioxide from the stack gases of advanced power plants would cost $35 to $264 per ton of carbon dioxide.[1] Removing carbon dioxide from existing plants could require an incremental electricity cost increase of about 50% according to a study reported in 1997.[2] But where can the carbon dioxide that is removed be discarded without causing an adverse environmental impact?

1. International Energy Agency. *"Carbon dioxide capture from power stations."* IEA, 1998.

2. Sacolow R. (Editor) *"Fuels decarbonization and carbon sequestration"* Report of a Workshop, PU/CEES Report Number 302, Princeton University, 1997

Equation 10.2 shows dissociation of methane to produce carbon and hydrogen. This process can be readily accomplished at about 900°C (1652°F) by heat addition of about 75.6 kJ/mole to release two moles of hydrogen. Solar concentrators can provide the heat needed to produce carbon and hydrogen from methane or other compounds that are now wastefully burned. No carbon dioxide is released. Figure 10.2 shows how the process can be automated.

Hydrogen can be utilized in ordinary engines that power distributed electricity generators. Heat cascaded from such engines into useful applications replaces customary combustion of fossil fuels for the same purpose. This enables overall energy utilization of more than 80% of the energy released by combustion of hydrogen compared to less than 40% by conventional central power plants. This efficiency gain far more than offsets the energy required for dissociation of hydrocarbons such as methane that have ordinarily been burned to produce heat.

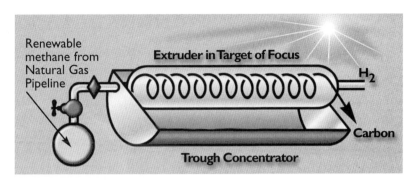

3. Steinberg M. *"Fossil fuel decarbonization technology for mitigating global warming"* International Journal of Hydrogen Energy, 24:771 (1999).

New Technology

In areas that do not have solar conditions suitable for concentration and endothermic processing operations, methane can be heated by combustion of as little as 10% of the incoming methane to release carbon and hydrogen from the remaining 90%.

Accomplishing conversion:

The most worthwhile collective effort in the history of Civilization is defined by the largeness of our present predicament. We must overcome global dependence upon burning the fossil equivalent of 200 million barrels of oil each day. Producing the will power needed to convert from dependence upon fossil fuels is the first step. Facilitating widespread participation in production, delivery and efficient utilization of hydrogen will follow as we make much better use of the existing infrastructure.

FACILITATING BUSINESS, NOT BUSINESS AS USUAL

International Renewable Resources Institute launched businesses will provide the market with business-development support. This is the key to job development and creation of a sustainable economy that improves the environment as it grows.

MYTHS

In order to be dangerous, myths must sell better than realities. Selling half-truths, impressions, and false hope is what myth marketing and "dis-information" specialists do best. The following may describe persons, places and incidents that are familiar. The intention is not to defame but to suggest better application of the enormous marketing and political talents now devoted to bogus promotions.

Myths about hydrogen have long caused confusion and delays. Myths have helped fraudsters deceive investors. Discussions about a few of the more interesting and infamous myths follow:

POLLUTION PORTENDS PROFIT: or burn fossils faster and make more money!

Myth 1.0

A modern myth of the Industrial Revolution is that pollution is a necessary circumstance of economic progress. One aspect of this myth suggests that Civilization should mine and burn more fossil fuels because doing so has been correlated for 150 years with increased economic activity and profit potential.

A N A L Y S I S :

Burning one gallon of oil forgoes the opportunity to produce over $35 worth of durable goods. Such durable goods can be designed for recycling into better inventions for future wealth expansion.

When a gallon of oil is refined and burned as fuel oil, gasoline or diesel fuel, the carbon in the fuel is converted into about 20 lbs of carbon dioxide along with other pollutants. These pollutants represent an enormous opportunity cost.

Oil can be converted into various polymers to produce computers, telephones, televisions, semiconductor devices, transportation components, clothing, paints, pigments, medical apparatus, and many other high-value products.

CHAPTER 11.0

PERPETUAL MOTION VEHICLES:
Eliminate any need for hydrogen suppliers!

This myth takes many forms and generally claims that you can convert a vehicle to operation on hydrogen and never need to refill with anything but water. A "one-up" variation of this myth claims that you do not have to refill with water because you can recover the amount of water needed from the exhaust of the engine or super fuel cell.

Conversion of a vehicle to this wonderful capability is supposedly achieved by installation of a special electrolyzer or "hydrolyser" that receives electricity from the vehicle's lead-acid battery. Hydrogen and oxygen are produced by passage of electrical current through water in the electrolyzer. The engine is supposed to operate adequately on the amount of hydrogen and/or oxygen produced by the electrolyzer.

Claims are based on the supposedly super efficient electrolyzer that splits water to make enough hydrogen to run the engine that powers the vehicle along with the electrolysis process. All this is to be accomplished from the electrical output of the vehicle's engine-alternator-battery system.

A N A L Y S I S :

Suppose the vehicle's engine is 30% efficient. (Even if the engine is 60% efficient, notice that the result is still a dead vehicle) If 100 energy units of hydrogen (or gasoline or any other fuel) are used to fuel this engine, it will only provide 30 energy units of shaft power. Even if all 30 energy units from the engine would be delivered to the alternator that might achieve a 90% efficiency, only 27 energy units of electricity are available for delivery to the battery or to the electrolyzer if the battery is by-passed.

If the super electrolyzer could achieve an amazing 100% efficiency, only 27 energy units of hydrogen would be available to continue engine operation. Obviously the engine will not continue to do the work that required 100 energy units with only 27 energy units so it slows down and after a while the engine would receive too little hydrogen to overcome internal friction and heat losses, then it would stop.

The problem is not the use of an on-board electrolyzer. It is the overzealous sales effort to have it do more than the laws of thermo-dynamics allow. The engine or fuel cell that required 100 units of fuel energy for step (1) will not continue to work well in step (2) with only 27 units of energy. In step (3) it will work less well because it receives only 7.2 units of energy. On the fourth time around (step 4 of this declining energy series) the engine would only receive 1.9 units of energy.

Myth 2.0

100 Units of Fuel Energy ① ② ③

30% Efficient Engine or Fuel Cell

70 Units of Heat Rejected

30 Units of Work

90% Efficient Alternator

3 Units of Heat

27 Units of Electricity

100% Efficient Electrolizer

27 Units of Hydrogen

② 27 units of energy

③ 7.2 units of energy

④ 1.9 units of energy

REGENERATIVE BRAKING:
One of the better applications of on-board electrolyzers.

A better way to utilize an on-board electrolyzer is to arrange for controlled conversion of the vehicle's kinetic energy into chemical fuel potential energy. In other words, use an on-board electrolyzer to make hydrogen instead of producing heat on the friction brakes at times that the vehicle is slowed. This has the potential of recovering half or more of the energy normally dissipated by the brakes when the vehicle is slowed. Electrolyzers can be designed to deliver pressurized inventories of hydrogen and oxygen for convenient storage. Engine operation on hydrogen or "Hy-Boost" mixtures of hydrogen and conventional fuels can be more efficient than on gasoline. It will not be the perpetual motion system that the sales department wants to sell, but using an on-board electrolyzer to convert stopping energy into hydrogen could improve vehicle range and fuel efficiency.

Some of the recently introduced "hybrid" vehicles utilize nickel metal-hydride batteries that provide electrolysis of water during regenerative breaking.

FUEL VAPORIZERS PROVIDE 100 MPG;
or vapor carburetors triple MPG so renewable hydrogen is not needed to reduce dependence upon oil.

Myth 3.0
Throughout most of the last century, there were reoccurring claims of phenomenal fuel economy achievements by carburetion techniques for vaporizing the gasoline before it entered the combustion chambers of an engine. Various approaches claimed to provide such benefits including:

1) Very fine wire mesh screens in the intake manifold which are supposed to break the fuel up into atomized or vaporous particles, and

2) Ultrasonic devices that are supposed to break the gasoline into vapors, and

3) Heaters such as exhaust heat exchangers or resistance heaters that are located somewhere in the intake system to heat the gasoline to form vapors.

A N A L Y S I S :

Assume that by some approach the gasoline (which is liquid at room temperature) is converted into vapor. What does this accomplish? Natural gas is mostly methane and it is gaseous at room temperature. Any benefit of a completely vaporous fuel can readily be found by testing the engine with natural gas. But either way, (with vaporized gasoline or natural gas) many of the results will be the same including:

1) The vaporized fuel will occupy more volume than the liquid state. This larger volume of fuel per energy unit will displace more air than the liquid form of the fuel and the engine will intake less air than with the same energy units of liquid fuel. With less air the engine's capacity for power production will be diminished with the vaporous form of fuel. Less air supports smaller fires, and less power is produced.

2) Once delivered into the combustion chamber, the vaporous
   form of fuel will require more compression energy than the
   liquid form of fuel as the piston closes the combustion chamber
   from BDC to TDC. This is measured as increased back-work.
   Some of this back-work may be recovered during the power
   stroke if the heat of compression has not been lost to the
   surrounding atmosphere by energy transfer through the piston,
   cylinder walls, and head. When using the vaporous form of fuel,
   it may be necessary to add flywheel inertia or increase the idle
   speed to overcome unacceptable variation in output shaft speed
   as a function of the compression-stroke deceleration and power-
   stroke acceleration.

Exhaustive testing on vaporized gasoline and vaporous fuels
such as methane and propane have shown that the fuel economy for
homogenous-charge engine operation at the same power output is at
best only marginally greater, under limited operating conditions, with
the vaporized fuel in comparison with liquid gasoline. This difference
is largely due to the energy consumed in the process of evaporating
the liquid droplets of gasoline to form vapor during compression and
the early stages of combustion. If this energy, called the latent heat of
evaporation, could be added from some waste energy source before
the fuel enters the combustion chamber it might be an advantage.
But the various claims of providing 100-mpg fuel economy for cars
that were previously achieving only 15 mpg to 30 mpg have never
been found in the many efforts that the author has made to
scientifically research various gasoline-vaporizing inventions.

More realistic than the 100-MPG claim for fuel-vaporizing
technologies are benefits such as reduced emissions of carbon monoxide
and unburned hydrocarbons during partially loaded engine conditions.
At higher power levels with homogeneous-charge operation, the engine
cannot get the air required to achieve full power without increased
emissions of carbon monoxide and unburned hydrocarbons.

NOTE: *WORLD FUEL ECONOMY RECORD:*
is not held by a vaporized homogeneous charge operation.

In further analysis of the fuel vaporization issue, it is very important to note that Diesel engines hold the world record for fuel efficiency by internal combustion engines. This record is not achieved by vaporous homogeneous-charge operating conditions! Diesel engines achieve higher fuel efficiency by directly injecting liquid diesel fuel into their combustion chambers. The heat produced by compressing the air and injecting the diesel fuel through one or more small holes must be sufficient to evaporate the liquid diesel fuel, thermally crack the fuel molecules, and to ignite the cracked fuel constituents.

Diesel engines hold the world record for fuel efficiency because of stratified-charge combustion within excess air. This is accomplished in spite of comparatively more backwork during compression, greater heat losses during compression, the requirement for intentional heat loss to evaporate the fuel within the combustion chamber, and incurred friction losses for the longer-stroke higher-compression ratios that are required.

Another similar myth that relates to fuel pretreatment claims to DOUBLE MPG WITH MAGNETS! The *Magic Magnet Sales Department* has an easier to install apparatus than the *Favorite Fuel Vaporizer Department*.

Ubiquitous ferromagnetic materials such as iron and various steels offer certain truly amazing properties. Think about the marvelous benefits that the magnetic compass has provided. A common steel nail can be made into a magnet by winding an insulated wire on it and establishing an electrical current in the wire winding. This will cause the nail to have aligned magnetic domains and exhibit strong magnetism when the current flows in the wire winding. The magnetized steel will pick up iron filings and attract other small iron-alloy objects that come near enough. Permanent magnets will do the same without the current in the wire winding. This strange attraction between magnetized and susceptible ferromagnetic objects is wondersome but unfortunately it does not provide more miles per gallon.

Even so it is a worthy brainteaser to try to explain how a solid piece of iron can exert force upon another piece of iron without touching it and make it do work. The energy that must be expended to accomplish the work done is equal to the magnetic force times the distance that the attracted masses move in response to the force.

EQUATION 11.1   Energy > Work = FD

Where does the energy come from to move magnetic materials? If a precision scale is used to weigh the magnets before and after they move, there is no perceptible change in their weight (mass). Their temperatures seem to stay nearly the same throughout the process. So what is the source of the expended energy? These mysterious and wonderful properties of magnets are used in electric motors and generators… and they work. So why not place some electromagnets or permanent magnets on your fuel line to do something beneficial for your fuel?

A N A L Y S I S :

The main problem is that they do not work as advertised! Magnets do not improve gasoline mileage because gasoline and other organic molecules are not magnetic and do not conduct electricity to become electromagnets and therefore do not respond to the presence of magnetic fields.

After testing many representations about 100 MPG carburetors and fuel conditioners that use magnets, it seems that no pattern or type of magnet makes any perceivable difference in fuel economy. The only advantage seems to be that the magnets may attract and keep unwanted steel filings in the fuel line from going downstream and clogging a fuel injector. But filters do this job more assuredly. Filters will remove unwanted sand, scale, Teflon tape debris and many other things that magnets will not attract.

SUGGESTION: The world's magnetic materials, particularly the rare earth metals that provide exceptionally strong permanent magnets, should be used in renewable energy applications and not wasted on misguided attempts to inspire gasoline to vaporize, homogenize, or pasteurize.

"WATER FUEL" PROVIDES COMPACT/CHEAP HYDROGEN GENERATOR:

Another promotion suggests that you can save space and money by putting water in your fuel tank and use it to make hydrogen by reaction with an active-metal. It is not the same as the perpetual motion promotion because this approach calls for replacement of the active metal and water that are depleted as hydrogen is produced.

Myth 4.0

A N A L Y S I S :

Water is composed of 8 mass units of oxygen per mass unit of hydrogen. Nine pounds of water has one pound of hydrogen that can be released by various reactions.

If you want to have as much hydrogen as needed to replace 128lbs (20 gallons) of gasoline, you need 40 pounds of hydrogen. This would require about 360 pounds of water if all of the hydrogen could be extracted. At 8.33 pounds of water per gallon, the gasoline tank would have to be stretched to accommodate at least 43.2 gallons of water to deliver 20 gallons of gasoline equivalent (GGE).

What is more difficult is the reactor to release the hydrogen from the water at the rate that it is needed. But, if the reaction releases the hydrogen all at once it would complicate things because adequate arrangements would be required to store over 7,000 cubic feet of hydrogen.

But an even greater problem is the energy source for releasing the hydrogen from water. It takes at least 61,050 BTUs of energy per pound of hydrogen. Regardless of the presence of any catalyst it takes at least 61,050 BTUs of energy to release the hydrogen from water. This energy must be in the form of heat and electricity for electrolysis, very high temperature heat for direct thermal dissociation, or some other form of energy that is appropriate for the process that is chosen for extracting hydrogen from water. Where is the energy source for releasing hydrogen and oxygen from water?

One possible approach is to take along a supply of active metal such as sodium, potassium, or calcium. A very vigorous reaction will occur with any of these active metals. But these reactions do not release all of the hydrogen, illustratively:

$$Na + H_2O \rightarrow NaOH + (1/2)H_2$$

EQUATION 11.2

In order to get the 40 pounds of hydrogen that is needed to replace 20 gallons of gasoline requires at least 720 pounds of water, 920 pounds of sodium, and some pretty special (and heavy) provisions for retaining the large amount (1600 lbs) of caustic sodium hydroxide that is produced. It is worse if potassium or calcium is used because these metals have molar weights of 39 and 40 compared to 23 for sodium.

So if someone tells you that they use water and an active metal reactant to make enough hydrogen to replace the gasoline originally used for the same range on some vehicle:

- Look for heavier duty tires, wheels, shock absorbers, springs, and serious reinforcement of the frame.

- Think twice about volunteering to have the residual caustic solution drained in your yard.

A greater yield of hydrogen could be achieved by a hydride such as sodium, lithium, boron, or beryllium hydride.

EQUATION 11.3 $NaH + H_2O \rightarrow NaOH + H_2$

Illustratively, compare sodium hydride with sodium. In order to get the 40 pounds of hydrogen that is needed, the process requires 360 pounds of water, 480 pounds of sodium hydride, and provisions for retaining the large amount (800 lbs) of sodium hydroxide that is produced.

OBSERVATION: But where should the sales department make their pitch? Stationary applications do not have the weight penalties of transportation systems. And there are other applications where small amounts of active metals or active metal hydrides work well.

HYDROGEN WEATHER BALLOONS:

Sodium and lithium hydrides are widely used as compact sources of hydrogen for filling weather balloons. One gram of lithium hydride releases 2.83 liters of hydrogen when reacted with water.

EQUATION 11.4 $LiH + H_2O \rightarrow LiOH + H_2$

HYDROGEN CAN'T BE CONTAINED: This myth claims that hydrogen seeps through steel tanks and other containers.

Myth 5.0

One of the most frequently discussed myths regarding hydrogen concerns the claim that it is impossible to keep hydrogen in captivity because it is the smallest molecule. A good authority such as a chemistry professor often repeats this myth.

A N A L Y S I S :

In 1964, I noticed a hydrostatic test date of "07" on a cylinder of hydrogen in my fuel cell laboratory at the University of Kansas. I asked a chemistry professor (who had stated that hydrogen was so small that it could pass through or embrittle most metals) what kind of liner was in the cylinder that was manufactured in 1907?

He said that he did not know and was surprised that the cylinder had been in commercial use delivering compressed hydrogen at 2,000 psig since 1907. Safety test dates starting in 1907 and every five years afterward were stamped on the neck area of the cylinder just beneath the safety cap that was threaded over the valve for protection during shipment.

The neck area seemed to be a safe area to remove a small amount of metal for analysis to determine what wonderful metal was able to contain hydrogen. From the center of one of the "7s" in a safety date, a few metal filings were carefully taken. Chemical analysis showed the metal to be a low-alloy carbon steel containing about 99% iron, 0.6% manganese, 0.2% carbon, and various other elements. Not as highly alloyed as a plow steel from the farm. This meant that there must be a very valuable liner material that was holding the hydrogen within the shell of low-alloy steel.

But careful inspection of the cylinder revealed that no liner was present. This low-alloy or "mild" steel cylinder had been in commercial service since 1907 providing countless trips to and from gas suppliers to welding shops, heat treaters, brazing houses, and university laboratories like mine. Eventually this cylinder emptied as I conducted tests on fuel cell catalysts and the "1907 cylinder" was routinely exchanged for a full cylinder of hydrogen, which happened to have a first safety test date of 1940. Later I received hydrogen that was delivered in a cylinder that was first tested in 1909 and I wondered if only old cylinders were able to hold hydrogen!

So I asked technical gas suppliers if old cylinders were better than new cylinders. The gas supplier confirmed that they had several pre-World-War I cylinders in stock along with numerous pre World-War II cylinders and that there was no reason to take a cylinder out of service so long as it passed scheduled safety checks.

After confirming the success of pre-World-War I cylinders in continuous commercial service, it occurred to me that low-alloy steel pipelines probably could carry hydrogen with the same efficiency and safety that the 1907 cylinder provided. I began an investigation of the suitability of low-alloy steels for pipeline distribution of hydrogen. All the theories seemed to indicate that the pipeline would do as well as the old cylinders were doing but when I asked pipeline authorities another myth was uncovered.

Pipeline authorities said that a common problem was pipe failures due to "hydrogen embrittlement" presumably from some source of tramp hydrogen in the pipeline. Deeper investigation showed that hydrogen embrittlement may indeed occur when the welding process used to connect lengths of pipe introduces a special form of hydrogen. Typical sources of hydrogen include dissociated water from damp welding rod flux or due to welding in a wet environment. The arc welding process electrolyzes water into hydrogen and oxygen ions. Ionic or nascent hydrogen particles are much smaller than normal diatomic hydrogen molecules. In the ionic state hydrogen is a proton or it joins water to form three protons and two electrons. In the nascent state hydrogen consists of a proton and an electron. In the molecular state hydrogen is two protons sharing two electrons.

Certified welders know how to avoid hydrogen embrittlement. They keep the welding rod dry in sealed canisters and prepare the welding area by cleaning it, preheating, and keeping the work pieces dry during the welding process. This avoids embrittlement by preventing hydrogen introduction into the steel.

Plating experts who use aqueous solutions to electroplate nickel, zinc, or chromium on steel also know that the plating operation introduces ionic or atomic hydrogen into the steel work piece.

After plating, the next step is baking the plated parts to remove the hydrogen before embrittlement causes weakening of the metal or rejection of the plating.

Several years ago I started a search for the oldest hydrogen cylinder with the help of local suppliers of hydrogen and specialty gases. So far, I have discovered two cylinders that were first tested in 1916 and two with 1917 indicated as the first year of service. At the end of World War I, Germany was forced to give up war materials, and the hydrogen cylinders were handed over to Allied victors as part of the war reparations program forced on Germany to keep Europe safe from another military adventure. Many of these cylinders were then re-branded like cattle after going to a new owner. Iron cross patterns were often changed to form an outer box with four smaller boxes within it.

One reason that hydrogen cylinders were taken from Germany at the end of World War I is because the hydrogen dirigibles had been used for observation posts, to hang anti-aircraft cables over strategic military locations, and for long-range flights. From about 1900 to 1937 German dirigible engineers such as Zeppelin and Von Hindenburg wanted to use helium as a lifting gas for lighter-than-air dirigibles but at that time America was the only source in the world of this rare gas.

Vast natural gas fields in the Hugoton, Kansas area produced small percentages of helium. U.S. engineers had learned how to extract the rare helium for uses ranging from electron tubes, to cover gas for welding (Heliarc welding), and for weather balloons. But the U.S. Government discouraged sales of helium to Germany after it became apparent that the German dirigibles could be used to gain the same military advantages that had been achieved by the North during the American Civil War. Germany was forced to use hydrogen for buoyancy in the dirigibles.

After World War II more hydrogen cylinders with the Swastika mark near the safety test date were handed over as part of the World War II reparations arrangement in Germany's surrender. If these hydrogen cylinders taken from Germany could talk they might tell of voyages from Berlin to and from New York or Rio de Janeiro, as the Zeppelins flew over the oceans at 75 miles per hour leaving ocean liners and short-range heavier-than-air planes far behind. These hydrogen Zeppelins offered far greater range than any other aircraft. They flew nonstop from Berlin to New York or Rio de Janerio. On board these lighter-than-air Zeppelins were cylinders of compressed hydrogen for re-filling the giant sausage skins that confined the hydrogen. Altitude was adjusted by the pressure and volume of hydrogen maintained in these animal-gut or latex casings that resembled giant sausage skins.

Hydrogen pipelines made of welded steel tubing are expanding. Refineries produce large amounts of hydrogen for making petrochemicals and for making cleaner gasoline and diesel fuels. Hydrogen is used to remove sulfur and to hydrogenate gasoline molecules to improve the hydrogen-to-carbon ratio which results in an improved octane rating.

An emerging trend is for industrial gas suppliers to produce and sell hydrogen to numerous refineries from pipelines. Some of the most extensive hydrogen pipelines are in Germany and the Houston, Texas area where the same hydrogen pipeline serves gasoline producers, ammonia synthesizers, and petrochemical plants. Natural gas pipelines bring a mixture of methane, and heavier molecules such as ethane, propane, and butane. The heavier molecules are increasingly used to make polymers and petrochemicals. Methane is reacted with steam to form methanol, carbon dioxide and hydrogen. The hydrogen is separated and added to the hydrogen pipeline.

CONCLUSION: 80-year old steel cylinders prove the long-term capability of steel for containment of hydrogen. Steel pipelines made of similar alloys have served as dependable hydrogen delivery systems. It is a myth that hydrogen cannot be efficiently and safely contained in steel containers and pipelines.

HYDROGEN EXPOSED TO AIR WILL INSTANTLY COMBUST: so don't be around it unless you want to be burned instantly.

This myth seems to be propagated by the semiconductor industry. At least it is often repeated by presumably well meaning technicians and engineers in the semiconductor industry. The erroneous belief is that hydrogen will ignite upon exposure to air if it is leaked from a hydrogen storage system.

Myth 6.0

A N A L Y S I S :

In fact, leaking hydrogen, like any other fuel, poses a danger. But, hydrogen requires an ignition temperature of 585°C (1085°F) to initiate combustion in ambient pressure air. Unless certain catalysts make it otherwise, the air and hydrogen must be mixed and heated to at least 585°C before the hydrogen will start to burn. If hydrogen is mixed with oxygen instead of air, the temperature of auto-ignition is reduced but a spark, catalyst, or heated surface is still required to ignite the mixture.

Regarding the related myth that hydrogen burns instantaneously if exposed to oxygen, for some 30 years the author has demonstrated electrolysis of water in which the hydrogen and oxygen are allowed to mix freely to form a stoichiometric mixture. None of these electrolysis demonstrations has caused a spontaneous hydrogen-oxygen fire.

In conclusion, what is required to initiate combustion of hydrogen in air at atmospheric pressure is:

1) An ignition source that heats a mixture of hydrogen and air to 585°C, or

2) A suitable catalyst such as platinum black, or

3) Ionizing radiation of sufficient intensity to start a sustained chemical reaction.

HYDROGEN IS POISONOUS:

Myth 7.0

A popular misrepresentation of the dangers that are associated with hydrogen is the claim that hydrogen is poisonous. This claim is sometimes made after reference to poisonous or otherwise dangerous compounds that contain hydrogen such as ammonia (NH_3), hydrogen sulfide (H_2S), hydrogen cyanide (HCN), and illuminating gas mixtures of carbon monoxide (CO) and hydrogen.

A N A L Y S I S :

Hydrogen is not poisonous. In fact, the first thing every green plant does in the photosynthesis process is to produce hydrogen from water. Humans also have natural exposure to hydrogen. Hydrogen can be measured to be up to 40% of human intestinal gases. Most people are unwittingly exposed to hydrogen on a regular basis without being poisoned, burned, or harmed in any other way.

The product of hydrogen combustion is water. Water is not poisonous. Airplane crashes have caused many deaths. A large percentage of these deaths occur because of poisonous fumes that are released by the partial combustion of jet fuel and various materials found in the passenger compartment.

CAUTION: CARBON MONOXIDE IS POISONOUS.

Death due to carbon monoxide poisoning can occur by breathing the poisonous exhaust of a vehicle burning gasoline or virtually any homogeneous-charge hydrocarbon fuel. Combustion of fuel-rich mixtures of natural gas, propane, gasoline, etc., causes carbon monoxide poisoning of persons who voluntarily or involuntarily breathe the exhaust.

STOICHIOMETRIC IMPLOSIONS: This myth claims that combustion of a stoichiometric mixture of hydrogen and oxygen is not dangerous!

Myth 8.0

Stoichiometric mixtures in the ratio of two volumes of hydrogen and one volume of oxygen produce water without any remaining hydrogen or oxygen. If water is electrolyzed, stoichiometric amounts of hydrogen and oxygen are produced. A very dangerous myth suggests that it is not dangerous to explode stoichiometric mixtures of hydrogen and oxygen. This myth is based on the supposition that if you burn a stoichiometric mixture of hydrogen and oxygen and instantaneously condense the steam that is produced, the resulting volume of liquid will be much less than that of the gaseous mixture before combustion. The reduction in volume is said to create an implosion and thus it is safe to "stoichiometrically implode" hydrogen and oxygen.

A N A L Y S I S :

Actually the combustion of one pound of hydrogen in a stoichiometric ratio with oxygen produces nine pounds of steam and releases 61,050 BTUs. If the process is adiabatic (no heat losses), the resulting steam temperature is a very hot 3,315°C (6000°F). At this temperature, the steam occupies many times more volume than the stoichiometric mixture of hydrogen and oxygen at room temperature.

$$H_2 + 1/2O_2 \rightarrow H_2O + 61,050 \text{ BTU}$$

EQUATION 11.5

What this myth misses is the important fact that 61,050 BTUs is a great amount of energy and until this much energy is converted to work or is transferred away from the steam that is produced, the steam will stay hot enough to do great harm to anyone who is exposed to it.

Another dangerous result of stoichiometric combustion is the sudden pressure rise. If this reaction is contained to prevent expansion, the resulting steam pressure can be very high until the heat released by combustion is removed.

It should also be noted that stoichiometric combustion produces the highest temperature compared to fuel-rich or fuel-lean combustion.

HYDROGEN COMBUSTION ELIMINATES RADIOACTIVE WASTES:

This myth claims that radioactive materials such as nuclear power plant wastes and certain hospital wastes will cease to be radioactive if they are exposed to a hydrogen flame.

Myth 9.0

A N A L Y S I S :

Radioactivity is not altered by hydrogen-aided combustion. The waste material may change in physical and chemical form but the radioactivity will remain. Radioactivity is due to conditions that arise from instability of the nucleus of the atom. Changing the electron configuration and/or chemical bond situation will not change the nucleus of the radioactive isotopes that are involved.

If a radioactive substance is on a cotton swab and the cotton swab is burned in a fire (any fire) the cotton can be converted to carbon dioxide and water vapor. The cotton may disappear as the carbon dioxide and water vapor escape into the atmosphere but the radioactive atoms will not be made into non-radioactive atoms by being exposed to the fire.

It is very dangerous to burn radioactive wastes without proper precautions because the radioactive material is more mobile than in the cool state before being exposed to the fire. The radioactivity will remain. Innocent believers of this myth place themselves at risk of dangerous exposures to radioactivate air pollution.

Breathing or otherwise ingesting a radioactive substance is very dangerous.

HYDROGEN IS MORE EXPENSIVE THAN GASOLINE. This myth is based on the fact that small amounts of high purity hydrogen that may be delivered occasionally from a specialty gas company are much more expensive than low-purity commodity supplies of gasoline from filling stations.

A N A L Y S I S :

Myth 10.0

At equivalent economies of scale, and if equally subsidized, hydrogen is much less expensive than gasoline. Most motorists know the apparent price of gasoline at the pump but few know the highest value application of the fossil resources represented or even the energy requirements or cost of producing gasoline on a sustainable basis. Gasoline and most of the fossil fuels that produce the electricity that the Industrial Revolution depends upon are virtually "stolen" (meaning that these fossil resources are taken from the Earth and sold at a price that is far below the replacement cost). Burning fossil hydrocarbons is stealing from ourselves and from future generations. To remedy this situation, investment in large-scale Renewable Energy Parks is needed to provide energy needed for sustainable prosperity without pollution.

Refineries separate various fuels from crude oil by distillation. Gasoline is evaporated from heated fossil hydrocarbons and selectively condensed to yield hundreds of molecular species that are blended to adjust the vapor pressure, octane rating, and various other characteristics. The idealized "average molecule" for gasoline is called octane (C_8H_{18}) which is composed of eight carbon atoms surrounded by eighteen hydrogen atoms. Octane is about 84.2% carbon and 15.8% hydrogen by weight.

A widespread myth supposes that the market price illustrates the expense of actually producing sustainable supplies of gasoline from water and air. Gasoline at $1 per gallon is priced far below the cost of sustainable production. Tax subsidies, depletion allowance, enhanced oil recovery credits, foreign tax credits, accelerated depreciation allowances, and military expenditures to assure oil deliveries to the U.S. along with associated health care and productivity losses amount to estimates between $5.00 and $15.00 per gallon of gasoline.

Ultimately, Earth's human population must find a way to progress past our dependence upon 60- to 600-million year old fossil hydrocarbons. Starting from sustainable ingredients such as sunlight, water, and air to make a gallon of gasoline requires an enormous endeavor and adds up to very large costs. If you hire a very clever organic chemist and have time to find and set up the equipment and make your next tank of gasoline from sunshine, water, and air the cost for 20 gallons will far exceed $1,000. Solar energy can be utilized to supply the enormous amount of energy needed to separate carbon dioxide from the air. Carbon dioxide is available at the volumetric concentration of about 0.03% in air but 72.7% of the CO_2 weight is oxygen which forms a very stable bond with carbon.

This means that for the 108 pounds of carbon needed to make 20 gallons of gasoline about 400 pounds of carbon dioxide must be separated from 900,000 pounds of air. This is a lot of air; it will occupy a space of about 12 million cubic feet. Getting the carbon that is needed will require very expensive air-separation and carbon-extraction equipment. It takes enormous amounts of energy to gather and extract carbon from air.

Depending upon how you approach the problem, sorting through 12 million cubic feet of air to find only three molecules of carbon dioxide in every ten thousand molecules of air can be a very energy-intensive process.

Then the carbon dioxide must be broken into carbon and oxygen.

Even if a perfect process could be devised, more than 75,960 BTUs would be required to separate enough carbon from carbon dioxide to make a gallon of gasoline. In other words, this is the most energy that will be provided by burning the 5.39 pounds of carbon needed to make a gallon of gasoline, but actual processes could be expected to expend three to ten times this amount of energy to separate carbon from carbon dioxide. Green plants gain carbon from the air for making plant tissues at about one-half of one percent (0.5%) efficiency. This low efficiency is barely sufficient to produce global food requirements and it is far too small for the world's large human population to depend upon for producing renewable gasoline.

Additional solar energy can be converted into electricity to electrolyze water to produce hydrogen. To provide the 1.01 pounds of hydrogen needed to make a gallon of gasoline will require at least 17.89KWH or 61,050 BTUs to dissociate water. It is possible to achieve 90% or higher electrolyzer efficiency. It is possible to operate an electrolyzer that produces hydrogen from water and directly pressurizes the hydrogen for compact storage without the additional requirement of a gas compressor. Even casual approaches usually achieve over 50% electrolyzer efficiency.

With carbon and hydrogen you can synthesize the various hydrocarbons that are needed to make gasoline. Gasoline is composed of hundreds of different hydrocarbons. Most of these hydrocarbons contain 4 to 12 carbon atoms per molecule. Gasoline requires special additives to stabilize it and the materials to which it is exposed. In order to solve problems with storage, performance, and emissions you will need to synthesize special additives to improve gasoline's storage stability, provide detergent action, improve octane, prevent polymerization, add oxygen, and inhibit rust formation during storage in steel tanks.

If you operate the complex air-sourced carbon extraction and gasoline production system on a continuous basis, the cost per gallon of manufactured gasoline and the required additive package may decline, but it will always be greater than making a simple substance like hydrogen from water in one clean step.

Illustratively, you can buy a wind generator or an array of photovoltaic panels, an electrolyzer, and use 44 gallons of water to produce enough hydrogen to replace 20 gallons of gasoline. Finding 44 gallons of water to make enough hydrogen to replace 20 gallons of gasoline is much easier than sorting through 12 million cubic feet of air to find and isolate 935 cubic feet of carbon dioxide, so you can expend much more time, energy, and investment on the carbon separation process, making hundreds of different types of hydrocarbon molecules and the additives needed in order to use gasoline as a fuel.

THE COST OF RENEWABLE HYDROGEN:

Producing the hydrogen needed to replace one gallon of gasoline will be much cleaner, safer, simpler, and far less expensive than making the complex organic blend of hydrocarbons in the gasoline recipe. And, as previously noted, electrolysis of water is much more efficient than separation of carbon dioxide from air and carbon from carbon dioxide. At an electrolyzer efficiency of 88%, about 40 KWH is needed to make enough hydrogen to replace a gallon of gasoline. A gallon of gasoline equivalent (GGE) of hydrogen provides, about 122,100 BTUs. The incremental cost of electricity (for off-peak interruptible production from established facilities) from hydroelectric plants is generally less than $0.005/KWH in Brazil, Paraguay, Canada, India, and China. Therefore, the electricity cost for producing hydrogen could be less than $0.21/GGE. If wheeling the electricity, electrolyzer amortization, and taxes brought the price to $1.00 per GGE it would be a positive step towards achieving prosperity without pollution with renewable hydrogen from a home refueling station.

Observations: Hydrogen has been stored at 2,000 PSI since the early 1900's in ordinary steel cylinders, which show no sign of degradation due to long-term exposure to pure hydrogen. Ordinary engines converted to operation on hydrogen show no sign of metal embrittlement or other degradation after fifteen years of pollution-free service. Engine oil stays clean, spark plugs last much longer, and degradation as measured by corrosion and wear on rings and bearings is greatly reduced.

HIGHEST VALUE USE OF FOSSIL HYDROCARBONS:

Regarding the highest-value use for fossil hydrocarbons, it is important to admit that the Industrial Revolution probably has not yet found the highest value for fossil hydrocarbons. However, it is imperative to note that 100 gallons of oil can be processed into durable goods that sell for over $3,500 including carpeting, fine clothing, TVs, CDs, electronic equipment, and transportation components.

Most of these durable goods are recyclable. Future generations can convert these recycled hydrocarbons into higher value products rather than disgracing our generation for foolishly burning the fossil feedstocks as fuels.

RECOMMENDATIONS:

Reproducing the hydrocarbons that are found in fossil reserves is energy intensive and expensive. The sustainable replacement cost of gasoline is much higher than present pump prices. It is much less expensive and more efficient to produce pollution-free hydrogen from water or agricultural wastes than it is to sustainably produce gasoline. Burning fossil hydrocarbons steals important opportunities to produce durable goods such as carpeting, TVs, CDs, and vehicle components. Future generations will need these recyclable materials as feedstocks for making higher value products.

In order to achieve large-scale production of renewable electricity and/or hydrogen it is imperative that investments on par with the U.S. highway system are provided to build needed plants and equipment. We must build Renewable Energy Parks that convert solar, wind, hydro, biomass and ocean energy into renewable electricity and hydrogen for achieving sustainable prosperity without pollution.

ADDING 50% WATER TO NAPHTHA OR GASOLINE WILL SOLVE THE OIL CRISIS:

It has been noted that mixtures of water and fossil fuels will burn, and therefore adding water to gasoline or naphtha will extend fossil fuel supplies. Other attempts to glorify water addition claim that as the fossil fuel burns it cracks or reacts with water to produce hydrogen, which then burns to release great amounts of heat.

Myth 11.0

ANALYSIS:

Adding water reduces the temperature of the combustion gases. Heat engine efficiency is dependent upon expanding the hottest gas that can be practically produced and contained. In order to heat water to the same temperature of combustion gases that are produced without water addition, more fuel must be burned.

In order to develop more power from an engine it is necessary to increase the average pressure in the combustion chamber. This can be done by adding more fuel and air by various techniques including turbocharging and supercharging. With more air delivery into the combustion chambers, more fuel can be burned and the engine will produce more power. However, the pistons may melt, or the valves may burn, or the rods may break if too much air and fuel are delivered to the combustion chamber.

In order to improve on power production in WWII fighters and bombers, it was a common practice to add water or a water-alcohol solution to the air-fuel mixture during takeoff. Adding water helped cool the air and made it denser so more air would be delivered into the combustion chambers. The same amount of turbocharger energy delivered more units of air because the air was cooler. With more air present, more fuel could be burned to boost power production.

Within the combustion chambers the water that first cooled the air continued to help by absorbing the heat of compression as droplets of water were converted into vapor. This reduced the back work during compression because the compression pressure was less.

This water evaporation role was also helpful in reducing the peak temperatures of pistons, valves and valve seats. Engines that were able to operate at steady state conditions at "X" power could operate during take-off conditions at 1.2 to 1.5 X power levels by adding 1.3 to 1.7 X gasoline.

In summary, adding water to a combustion engine can:

1) cool the air to allow increased air intake,

2) provide reduced pumping losses during compression, and

3) reduce peak combustion temperature.

Adding water does not add energy. After credit is taken for the use of water to cool incoming air and to decease the temperature and pressure during compression, it is still necessary to burn more gasoline to provide more power.

The gasoline or diesel fuel remains a fossil fuel and at best, the addition of water provides more air with less expenditure of pumping energy. Depression of the peak combustion temperature can help reduce nitrogen monoxide formation. Adding water with gasoline as a homogeneous charge with air requires throttling the air to achieve combustible air-fuel ratios during city driving conditions.

OBSERVATIONS: Greater benefits including reduced pumping losses, prevention of nitrogen monoxide, improved combustion rates, reduced carbon monoxide, and reduced hydrocarbon emissions can be achieved by unthrottled air entry followed by direct injection with positive ignition to achieve stratified-charge combustion.

COLD FUSION WILL PROVIDE ELECTRICITY THAT IS TOO CHEAP TO METER (Which sounds like the mythical "too-cheap-to-meter" nuclear power plant chant.)

Myth
12.0

Cold fusion is envisioned as a process for converting a very small amount of matter into energy as hydrogen undergoes mysterious cold conversion fusion into helium. It was claimed that a certain amount of energy could be multiplied into a larger amount of energy. Following the announcements by Fleshman and Pons about achieving "COLD FUSION," promoters and Charlatans jumped into action to gather much gullible money which poured in from various origins including state and federal sources.

A N A L Y S I S :

One problem with cold fusion is that it is cold. Even if cold fusion could "work" like hopeful promoters say it does, the problem of being "cold" remains. If Sadi Carnot were alive today, he might have something to say about the "cold fusion" designation. He might have suggested the name of the proposed phenomena as "warm" or "small fusion" in hopes that it could be as efficient as solar concentrators for providing high temperature heat for engines that operate through a temperature difference.

Application of Cold Fusion in practical engines would have to start with generation of heat that would be transferred through the walls of a heat exchanger. Generally, therefore, the temperature of the heat exchanger material would limit the Carnot efficiency.

What is called "cold fusion" seems to start with electrolysis of water, and after a while an exotherm is said to occur that causes the water to boil. If the heat released to boil the water is in fact greater than all of the energy used to purify the water, mine and manufacture the electrodes, and electrolyze the water; the difficult problems of thermodynamic quality and energy availability remain.

The cold fusion process proposes to take a high availability form of energy (electricity) and make a much lower availability form of heat energy.

WOULD DOUBLING COLD FUSION ENERGY BE ENOUGH?

Suppose a cold fusion process starts with X amount of electricity and could continuously release 2X amount of heat at 300°F which is considerably above the boiling temperature of ambient pressure water. If the use of the heat was by some ingenious and thermodynamically perfect device that could reject heat at room temperature, (80°F or 540°R) the Carnot limit for converting the 2X amount of heat into shaft work is:

$$W = \frac{2X\,(T_H - T_L)}{T_H} = 2X \left(\frac{760°R - 540°R}{760°R}\right) = 0.58X \qquad \text{EQUATION 11.6}$$

(But this is an overall loss of compared to the 2X energy addition)

At about 620°F the process could theoretically break even, but real-world machinery such as complex power plants generally achieve only about 30% to 70% of the potential Carnot efficiency, so the "cold" temperature that is needed rises significantly. At 1,435°F, which is a much higher temperature than modern steam power plants endure, the "double your energy" cold fusion process might barely be sustainable.

$$W = 2X\,(1354°R/1894°R) = 1.43X \text{ (Theoretical) and} \qquad \text{EQUATION 11.7}$$
$$\text{about 1X (Probable).}$$

So, why take X amount of electricity and put it into a complicated, expensive, and potentially dangerous cold-fusion power plant (that has to operate at a higher temperature than most modern steam power plants) to receive about the same amount of electricity that you had when you started the cold fusion process?

It would be much better to use the X amount of energy to build modular Solar Dish Gensets that use the clean, dependable, natural, well-proven, EPA and OSHA approved hot-fusion radiation from the Sun. Solar Dish Gensets have provided over 100 million watt hours of electricity which has been delivered to customers on the grid at an annual efficiency of approximately 23% for the process of converting free solar energy into electricity. This safety takes very high availability solar heat through the Stirling or Ericsson heat engine process of making mechanical work to produce electricity.

If you live in a wind-rich area take 1X energy to produce a wind generator. The resources needed to manufacture and operate wind generators or Solar Dish Gensets are much less than what is proposed for cold fusion power plants and best of all, we know that wind generators and Solar Dish Gensets work. The sun provides enormous amounts of clean radiation to earth. Every day the amount of solar energy reaching Earth is equivalent to 560 billion tons of coal. It is from natural, stable, big, safe, hot fusion; 93 million miles away. And the Sun will last another 4 or 5 billion years.

ELECTRIC CARS POLLUTE LESS THAN H$_2$ ICE VEHICLES

Myth 13.0

This myth is based on the correct observation that the vehicular emissions from an electric car are much less than from a gasoline car and consist (potentially) of some battery fumes and ozone from the arcs and sparks of control switches and brushes. So far, so good; and we all cheer the progress that electric propulsion systems are making.

A N A L Y S I S :

Minus emissions are better than zero emissions!

Ordinary Internal Combustion Engine (ICE) can be converted to produce full power on stratified-charge hydrogen within unthrottled air. In this mode, it is relatively easy to control the excess-air combustion process to prevent the peak flame temperature from exceeding about 2,200°C (4,000°F), and thus avoid the formation of nitrogen monoxide. And, there are no emissions of carbon dioxide, particulates, sulfur dioxide, hydrocarbons, or carbon monoxide. Even better than no emissions except water vapor, in most cities the stratified charge hydrogen ICE will actually clean the air that passes through the engine to accomplish "minus emissions" operation.

As previously noted, carbon dioxide is the dominant constituent of the greenhouse gas problem and causes about 60% of the atmospheric heat-trapping effect. Now, consider what happens in the process of producing the electricity that is needed to charge the batteries of the electric car. Most of the electricity in the U.S. is produced by burning coal in power plants. Somewhat less electricity is produced by use of natural gas, oil and nuclear fuels. The carbon dioxide emissions produced by the electric power industry already exceed the carbon dioxide emissions from all the gasoline and diesel fueled cars and trucks in the U.S.

ICE fuels such as hydrogen and methane can be produced from landfill gases and delivered to retail stations at a lower cost than producing gasoline from the same renewable feedstocks. Using these renewable energy fuels in transportation applications greatly reduces pollution to the atmosphere by displacing fossil gasoline and by preventing the methane from rotting into the atmosphere. Biomass sourced fuels can power an ICE with great cost-reduction, pollution abatement, and national security benefits. The overall contributions of carbon dioxide, acid-forming gases, and particulates from an ICE vehicle using landfill or agricultural hydrogen and/or methane are greatly reduced compared to the electric car with battery charging by the existing mix of electric power plants.

Enormous amounts of landfill and agricultural wastes rot into the atmosphere. These renewable resources should be converted into supplies of hydrogen and methane and used as cost-effective fuels to displace gasoline and diesel fuel. The amount of carbon that could be extracted from such wastes exceeds the annual production of coal in the U.S. This renewable carbon may be utilized to produce valuable goods such as carbon-reinforced golf clubs, tennis rackets, and vehicle components and aircraft materials..

Tables 10.1 and 10.2 are shown again for reader convenience.

TABLE 10.1

COMPARISONS OF GREENHOUSE GAS IMPACT

SPECIES	CONCENTRATION (ppbv)*	RATE OF INCREASE (% PER YEAR)	CONTRIBUTION (Relative % of TOTAL)
CO_2	353×10^3	0.5	60
CH_4	1.7×10^3	1.0	15
N_2O	310	0.2	5
CFC-12	0.48	4.0	8

*(ppvb) = Parts Per Billion By Volume

TABLE 10.2

HEAT TRAPPING CAPACITY

ATMOSPHERIC SPECIES	RELATIVE HEAT TRAPPING EFFECT	DECAY TIME (YEARS)
CO_2	1	120
CH_4	70	10
N_2O	200	150
CFC-12	6000	120

Reference: Henning Rodhe, A comparison of the Contribution of Various Gases the Greenhouse Effect, Science, Vol 24-8, June 1990.

THE HINDENBURG FIRE IN 1937 and the SPACE SHUTTLE DISASTERS in 1986 and 2003 PROVE THAT HYDROGEN IS TOO DANGEROUS FOR THE PUBLIC TO USE.

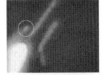

Sequence that lead to the Challenger disaster.

Myth 14.0 The Hindenburg fire in 1937, the Challenger's solid-booster O-ring failure in 1986, and the Columbia's re-entry breakup in 2003 are often mentioned to suggest that hydrogen is too dangerous for public use.

A N A L Y S I S :

The Hindenburg would have burned and crashed even if it had been using helium as a lifting gas. The Challenger failed because an inadequate O-ring seal in the solid booster rocket leaked. Columbia broke up on re-entry to the Earth's atmosphere.

During the American Civil War more than 3,000 hydrogen balloon flights occurred in a war zone without a single balloon fire. From 1900 to 1937, Hindenburg style of rigid "airships" traveled around the world without loss of life due to hydrogen caused fire. Graf Zeppelin, a smaller hydrogen airship, had made 650 flights. More than 18,000 passengers were delivered safely during the nine years that the Graf Zeppelin flew. This dirigible traveled more than one million miles (equivalent to 40 times around the world) and 144 flights were nonstop to and from Berlin across the Atlantic to Rio de Janeiro or the New York area.

The Hindenburg had crossed the Atlantic 21 times and had a fully loaded range of about 10,000 miles or about 5 to 6 days at cruise speed. The fateful fire that destroyed the Hindenburg started on the aluminum powder filled surface coating that covered this giant airship. The fire started in the outer skin near the tail section as the Hindenburg attempted to land during a rainy evening. An electrical engineer and crew member, Otto Beyersdorff wrote an investigative report on June 28, 1937 indicating the cause of the disaster: "The actual cause of the fire was the extreme easy flammability of the covering material brought about by discharges of an electrostatic nature ..."

Recently, NASA investigator Dr. Addison Bain independently verified this finding by scientific experiments that duplicated the vigorous ignition by static discharge to the aluminum powder filled covering material.

Investigation of the Hindenburg disaster proved that it was powdered aluminum in the flammable paint varnish that coated the infamous airship, not the hydrogen that started the fateful fire.

Since 1981, Space Shuttles have traveled far faster, carried larger and heavier payloads to space, and traveled more miles than any other type of vehicle. The Challenger and Columbia tragedies were not caused by hydrogen.

IF VEHICLES USE HYDROGEN, CITY STREETS WILL BE FLOODED WITH WATER FROM THEIR TAIL PIPES.

This objection to progress is based on the accurate observation that hydrogen produces water when it is burned in an engine or used in a fuel cell. And the erroneous conclusion that substituting hydrogen for gasoline would cause city streets to be flooded with water condensed from the tail pipes of cars that use hydrogen.

Myth 15.0

A N A L Y S I S :

Actually, using renewable hydrogen greatly reduces the net amount of water compared to the volume that is released by burning gasoline. And, please note there are even larger benefits for the Solar Hydrogen Civilization.

Gasoline is composed of approximately one thousand different molecular types that have an average carbon to hydrogen ratio of 2.25:1 as in octane or C_8H_{18}. Equations 11.7, 11.8, and 10.9 show the stoichiometric combustion of various amounts of C_8H_{18}.

$$C_8H_{18} + 12.5O_2 \rightarrow 8CO_2 + 9H_2O$$

EQUATION 11.8

$$114 \text{ lbs} + 400 \text{ lbs} \rightarrow 352 \text{ lbs} + 162 \text{ lbs}$$

EQUATION 11.9

$$6.4 \text{ lbs/gallon} \rightarrow 19.76 \text{ lbs } CO_2 + 9.09 \text{ lbs water}$$

EQUATION 11.10

One gallon of gasoline weighs about 6.4 pounds; therefore as summarized by Equation 11.9, combustion of 6.4 pounds of gasoline produces about 19.76 pounds of carbon dioxide and 9.09 pounds of water. One gallon of water weighs about 8.33 pounds; therefore burning one gallon of gasoline produces more than a gallon of water which passes out of the tail pipe, usually in vaporous form.

Water produced by burning fossil fuels that were stored in deep geological formations for 60 to 600 million years is being added to the Earth's surface inventory of water at the rate equivalent to about 190 million barrels per day. Much of this additional water is exhausted from vehicles that use fossil fuels as they are driven on city streets. On cool days you can see carbonic acid that forms as carbon dioxide is absorbed into the water that drips from the tail pipes of vehicles that use gasoline. Burning one gallon of fossil gasoline produces more than one gallon of condensable water from the exhaust.

Renewable Hydrogen Makes Zero Water Addition:

In comparison, consider the use of renewable hydrogen in vehicles and for electric power production. Replacing one gallon of gasoline will be accomplished by two pounds of hydrogen, which can be produced from 18 pounds of water. Burning the renewable hydrogen in an automobile or some other energy conversion operation will return the 18 pounds of water that was used to produce it. The net effect is zero water addition because the same amount of water that sourced the hydrogen is released when it is combusted or utilized in a hydrogen fuel cell.

AN ENCYCLOPEDIA OF NEW AND BETTER CARBON PRODUCTS, INSTEAD OF POLLUTION

When renewable hydrogen and carbon are produced from garbage, sewage, and agricultural wastes, even larger reductions in environmental impact and related benefits are achieved. Organic wastes can be collected and converted into renewable hydrogen, carbon and soil nutrients. Health-care costs are eliminated or greatly reduced for treating diseases due to airborne pollutants.

Producing durable goods from the carbon and using two pounds of hydrogen to replace each gallon of gasoline, diesel fuel, or fuel oil will reduce the carbon dioxide impact now incurred by burning fossil hydrocarbons. Burning a gallon of gasoline, diesel fuel, or fuel oil releases about 20 pounds of carbon dioxide at the point of use along with additional amounts that are released to refine and transport these fuels to market. Using renewable hydrogen will have zero net impact on the net water inventory and can reduce global carbon dioxide releases from fossil fuel combustion by about 158,000,000,000 pounds each day.

In addition to reducing the daily carbon dioxide emissions by 158 billion pounds, the process of using hydrogen that is derived from seawater, methane hydrates, sewage water or other contaminated water sources will enable collection of purified water from the exhaust systems of engines or fuel cells that use the hydrogen. The daily gain of good water from engines and fuel cells would be over 17 billion gallons per day.

In other words, 17 billion gallons of good water will be produced each day from otherwise objectionable water sources. This is more than two gallons of potentially collectable water per day for each person on Earth as a bonus for not polluting the air with 158 billion pounds of carbon dioxide.

A far-reaching additional benefit and economic driver will be production of durable goods from 40 billion pounds of carbon per day instead of allowing the carbon in fossil fuels to form 158 billion pounds of greenhouse gases.

IF ENERGY PRICES INCREASE, DISCOVERY EFFORTS WILL INCREASE, PRODUCTION AND REFINING PRACTICES WILL IMPROVE, AND MORE CHEAP COAL, OIL, AND NATURAL GAS WILL FLOW TO THE MARKET.

However, we cannot afford another expensive "Manhattan-type" project to build renewable energy parks or take time to explore space for another planet with exploitable fossil resources. We are financially strapped to pay for the world's military forces to maintain status quo in the global pecking order. We cannot think of spending the enormous amount needed to change from fossil fuels to renewable energy sources because we already have record national debts throughout the world even though energy is cheap.

Myth 16.0

This misleading myth supports the type of illogical rhetoric noted in the previous paragraph. However, analysis shows that for 150 years, higher prices have stimulated steadily greater oil production. So why worry about Peak Oil. Why not stay hooked on oil and be happy with Peak Profits for those in control of Peak Oil!

A N A L Y S I S :

Higher prices cannot endlessly spur more fossil fuel production. Like all other matters that depend upon stealing, the fossil economy is a temporary economy. Civilization cannot afford to wait for Earth's fossil reserves to become more depleted before investing sufficiently in ways and means to harness renewable resources to overcome dependence upon fossil and nuclear energy. It will be less expensive and far more civilized to convert the global economy to renewable energy than to continue to subsidize increasingly harmful dependence upon fossil resources.

Throughout this book, mention of "stealing" resources from the Earth has implied that the economic analysis of the practice of mining and burning fossil substances has established an ultimately harmful result. Harm, in this instance, comes from the erroneous belief that the value of the fossil resource is the same as the taking cost and that price increases will always cause more fossil resources to materialize. Ignoring the replacement costs of energy resources has allowed Civilization to operate with dangerously misleading beliefs about the economy.

OBSERVATIONS: Stars will still shine and the processes of the Universe will not be changed by Civilization's failure or success in overcoming the dilemma of dependence upon annually burning over one million years' of fossil accumulations each year. Human nature however will surely respond more favorably to overcoming the dilemma by seeking larger returns on the enormous investments that have been made to build the fossil-dependent economy, than if leaders advocate scrapping the engines, transportation equipment, factories, pipelines, electric grids, and farms that now depend upon fossil energy. A much better outcome will be accomplished by better utilization of the enormous infrastructure and technologies that have been developed by the temporary fossil economy for the purpose of launching the Renewable Resources Revolution.

Any program to do so that depends upon a central government's subsidy or grant will probably cost more than if grass-roots entrepreneurs find ways to launch the Renewable Resources Revolution. The International Renewable Resources Institute, Clean-Air Lotteries, and utilization of government lands in leases to renewable resources enterprises are suggested as ways for communities to gainfully compete in the Race to Save Civilization.

Communities can send delegates to the International Renewable Resources Institute for technology transfers and to develop business plans that successfully compete in the world market for capital. Communities with clean air and water will attract more new businesses along with more tourists and vacationers. More investor confidence in needed new ventures to expand energy independence will follow as communities seek higher living standards. Local non-profit administration of Clean-Air Lotteries can be designed to be much more efficient and faster delivery of clean air and local energy independence than waiting for federal taxes to be sufficiently allocated to attempt the same transition by bureaucratic central planning and political processes.

Local, state, and federal governments could do well to guarantee bonds that finance new ventures to make, use, and sell goods and services that provide energy independence. Local and national governments could provide tax and regulatory incentives for development of renewable resources. In the USA, federal and state governments could also allow renewable energy ventures to participate in fair competition by elimination of the $100 billion in direct and indirect annual subsidies that encourage depletion of fossil and nuclear fuels.

Improved return on the investment has been made in engines by converting to hydrogen (shown on table 4.5 on page 56). Existing engines now applied to transportation, farming, and electricity generation applications can be retrofitted to use hydrogen and clean the air, last longer, and produce more power. Conversion of such engines to interchangeable operation on hydrogen or petrol will create the market pull that is needed for entrepreneurs to launch hydrogen production operations and filling stations. Interchangeable operation on petrol will enable vehicles to smoothly return to fossil fuels if they need to travel outside of emerging hydrogen corridors.

CONCLUSION:

During its 150-year reign, the "Petroleum Age" has never met the needs of the larger portion of Civilization's rapidly growing population. Billions of families do not have sufficient fossil energy to earn enough to provide adequate food, water or housing. It is time for the Renewable Resources Revolution to end "Oil Wars" and provide opportunities throughout the world for much higher living standards. Widely distributed solar, wind, wave, and biomass resources are entirely adequate for achievement of sustainable prosperity. Forever!

THE COMING HYDROGEN ECONOMY POSES A GREAT THREAT TO PROTECTIVE OZONE IN THE STRATOSPHERE

This myth is based on the observation that hydrogen (H_2) is smaller than methane (CH_4) and would leak faster from the same leakage paths that now allow methane to contaminate the atmosphere. This myth is supported by observations that there are vast numbers of natural gas wells, countless miles of pipelines, and multitudes of cryogenic tanker deliveries required to produce and deliver the enormous amounts of natural gas being marketed. And that such commercial activities correlate to substantial annual leakage of methane and with the build up of methane in Earth's atmosphere.

Myth 17.0

A N A L Y S I S :

Actually the Solar Hydrogen Economy will greatly reduce destruction of protective ozone in the stratosphere. It will do so by prioritizing conversion of fugitive methane into sequestered carbon for producing durable goods and hydrogen for energy conversion purposes. It is also important to note that halogens are the primary cause of ozone destruction in the stratosphere and that halogens can be safely removed from the stratosphere and safely precipitated as salts into the oceans following reactions with atomized sodium and/or sodium hydroxide.

Atmospheric concentrations of CH_4 before about 1750 A.D. ranged from 676 to 716 ppb (parts per billion). After 1750 A.D., concentrations of methane increased more than 100% to the present level of about 1700 ppb. Anaerobic decay of organic wastes, termites, and soil erosion contributes more methane to the atmosphere than natural gas pipeline leaks. Human activities release 34.6% of the total atmospheric methane from anaerobic landfills and 26% from livestock farming. This compares to about 19.7% from natural gas production and distribution and 10% from coal mining.

Another source of methane is from anaerobic decay of organic components of eroded soil and sea organisms that fall to the ocean floor. Methane hydrates formed in the anaerobic ooze of the ocean floor contains more than two-times as much carbon as all coal, oil, and natural gas reserves on the contintents.[1]

Methane (CH_4) is composed of one carbon atom and four hydrogen atoms. Reactions that destroy ozone by oxidation of the carbon and hydrogen delivered by methane to the stratosphere are summarized as follows.

$$CH_4 + 4/3 O_3 \rightarrow CO_2 + 2H_2O \qquad \text{EQUATION 11.11}$$
$$CH_4 + 4 O_3 \rightarrow CO_2 + 2H_2O + 4O_2 \qquad \text{EQUATION 11.12}$$

One molecule of methane destroys 4/3 or 1.33 molecules of ozone in the best (but least probable case) and 4 molecules in the instance that is summarized in Equation 11.12.

However, considerably greater destruction of stratospheric ozone is caused by halogenated hydrocarbons. Particularly harmful agents of ozone destruction are manmade compounds known as halocarbons that are sufficiently inert to avoid entry into chemical reactions in the lower atmosphere. Such halogenated molecules reach the stratosphere in Earth's constantly moving and mixing atmosphere. Once delivered they readily enter into reactions with ozone and/or become dissociated by ultraviolet radiation to release halogens such as chlorine and bromine and cause virtually endless destruction of ozone.

Chlorine and bromine continue to cause ozone destruction without end. Each atom of chlorine (or bromine) that reaches the stratosphere is estimated to cause destruction of 100,000 ozone molecules before these serial killers are somehow randomly removed to the lower atmosphere.

In addition to CFC-12 (CCl_2F_2) which represents about 32% of the halocarbon molecules that reach the stratosphere, CFC-11 (CCl_3F) represents 23%, methyl chloride (CH_3Cl) represents 16%, carbon tetrachloride (CCl_4) comprises 12% and CFC-113 (CCl_2FCClF_2) adds about 7%.

About 3,000 times greater methane concentration than halogenated molecules presently exists in the global atmosphere. However each chlorine or bromine atom derived from a halogenated molecule that reaches the stratosphere will probably destroy 100,000 times more ozone than each molecule of methane.

Chlorine is about 170 times more prevalent in the stratosphere than bromine.[4] The general reactions by which halogens such as chlorine and/or bromine destroy ozone are summarized below.

EQUATION 11.13 $\qquad Cl + O_3 \rightarrow + ClO + O_2$

EQUATION 11.14 $\qquad ClO + O \rightarrow Cl + O_2$

Thus once chlorine or bromine enters the stratosphere these atoms cause an endless chain of ozone destroying reactions. This process is often said to be "catalytic" because the culprit chlorine and/or bromine atoms are not consumed by the reactions that destroy ozone. The net result of the catalytic destruction of stratospheric ozone by halogens is:

EQUATION 11.15 $\qquad O + O_3 \rightarrow 2O_2$

Catalytic destruction of ozone is also caused by hydrogen (H_2) that reaches the stratosphere to destroy ozone by producing a water molecule. High-energy ultraviolet radiation dissociates the water molecule to release a hydrogen atom that continues the ozone destruction. (But this is equally true of water formed by the reactions of Equation 11.11 and 11.12.) However, compared to methane a water molecule requires considerably more energy (shorter wavelength radiation) for dissociation. Therefore the probable initiation rate for ozone destruction in the stratosphere by methane is greater for equal molecular concentrations of hydrogen and methane. Further, each molecule of methane delivers four hydrogen atoms for the potential catalytic process compared to two hydrogen atoms that are delivered by a water molecule or by diatomic hydrogen.

Earth continually receives hydrogen from space and looses hydrogen that escapes into space. Earth is continually pelted with ice particles that evaporate upon colliding with the upper atmosphere. Diatomic hydrogen is present in space that Earth encounters as we orbit the Sun and such hydrogen has persisted through the eons of exposure to hard radiation. The chemical process equilibrium in space strongly favors maintenance of diatomic hydrogen.[5]

Through the eons of Earth history, water vapor has been constantly supplied by the oceans to interact with stratospheric ozone and hydrogen has been constantly supplied from space. Human activities that release chemicals including halocarbons and methane have changed this equilibrium.

In comparison with four molecules of ozone destruction by a molecule of methane only one molecule of ozone is consumed by a molecule of hydrogen that reaches the stratosphere. This is summarized by Equation 11.16

$$H_2 + O_3 \rightarrow + H_2O + O_2$$ **EQUATION 11.16**

Hydrogen is a much better choice for energy storage and conversion purposes than hydrocarbons in comparisons of greenhouse gas and ozone destruction hazards.

Regarding leakage of hydrogen from containment such as a tank, valve or pipeline, it is important to note that for an equivalent high-pressure drop through the same size crack or orifice, about 2.8 times as many hydrogen molecules would leak compared to methane. However the amount of stratospheric ozone destruction would be less with hydrogen than because of methane. This is because each molecule of methane can destroy four molecules of ozone but each molecule of hydrogen will probably destroy one molecule of ozone during the same time.

PROFITS FOR PREVENTING METHANE PROBLEMS:

The Solar Hydrogen Civilization will profit by reducing the amount of methane that reaches the stratosphere by sequestration of carbon for production of durable goods. This will be accomplished by using renewable energy such as concentrated solar energy to dissociate methane into carbon and hydrogen.

Landfills, livestock operations, the natural gas industry and coal mining sources of methane can profit by sequestering carbon and producing hydrogen. A high priority should be established for conversion of methane hydrates to carbon products and hydrogen.

Past negligence regarding methane addition to the atmosphere from landfills, livestock operations, natural gas production activities, and coal mines followed the economic analysis that started with the "taking cost" as the value of methane. It was cheaper to discard vast amounts of low-pressure or hydrogen-sulfide contaminated methane than to collect, purify and pressurize it compared to taking it from high-pressure wells that produced pipeline quality natural gas.

Achievement of global prosperity will be accomplished by emphasizing use of carbon-enhanced products that are lighter, stronger, and more corrosion resistant than conventional materials. Profits will be earned from sales of such durable goods that are made from sequestered carbon produced from methane that can be readily taken from sources that now contaminate the atmosphere. Accounting practices and economic analysis that values the replacement cost and not the taking cost will go far in establishing hydrogen production, containment, and delivery systems that leak far less than found in past practices of the waste disposal, natural gas and coal mining industries.

USING H$_2$ TO REMOVE HALOGENS THAT REACH THE STRATOSPHERE:

Chlorine, bromine and other halogens can be safely removed from the stratosphere and precipitated into the oceans following reactions with atomized sodium and/or sodium hydroxide. Sodium, hydrogen and oxygen derived from seawater will facilitate this remedial removal of halogens from the stratosphere.

Mixtures of hydrogen and oxygen can provide the non-polluting propellant for naval guns that are converted into sodium launchers. Large bore naval guns can be converted from previous use for delivering explosive shells to defense of the environment by utilizing mixtures of hydrogen and oxygen to launch sodium into the upper atmosphere for purposes of reacting with halogens to form salts that precipitate into the oceans.[6]

In other instances, hydrogen filled balloons will lift sodium payloads to the stratosphere and support solar concentrators that atomize sodium for the precipitation reactions.[6] If needed, these reactions can be shielded from radiation by aluminized solar-sail films. In both approaches for utilizing hydrogen to deliver sodium to the stratosphere, the remedial reactions are summarized in Equation 11.17 and 11.18.

EQUATION 11.17 $Na + Cl \rightarrow NaCl$

EQUATION 11.18 $Na + Br \rightarrow NaBr$

Therefore, the *solar hydrogen economy* provides practical ways to create a wealth-expansion economy while virtually eliminating destruction of protective ozone in the stratosphere due to reactions with manmade chemicals. The solar hydrogen economy will greatly depress emissions of greenhouse gases to reduce weather extremes including increased incidence and severity of hurricanes, tornados, lighting strikes, floods and mudslides. This can be accomplished by prioritizing conversion of biomass, gas hydrates and fossil sourced methane into valuable carbon durable goods and hydrogen for energy conversion purposes. **Each stratospheric ozone rescue** mission should be named after an outstanding volunteer or organization that has dedicated time and effort to advancement of *The Solar Hydrogen Civilization.*

References:

1. William P. Dillon, USGS Report and *"Gas Hydrates in the Ocean Environment"* Encyclopedia of Physical Science and Technology, Third Edition, Volume 6.

2. Rodhe Henning; *"A Comparison of the Contribution of Various Gases in the Greenhouse Effect"* Science Vol. 24-8, June 1990.

3. Ontario Energy Educators (1993).

4. U.S. EPA web references regarding ozone depletion: www.epa.gov/ozone/science/process

5. Paul Marmet, *"Discovery of H_2 In Space Explains Dark Matter and Redshift"* 21st CENTURY Science and Technology, Spring 2000, Pages 5-7.

6. Figure 8.1

WHAT WILL HAPPEN...
How Does it End?

Virtually everyone wants to know how it will end. Some readers cannot resist the temptation and turn to the end of a book to learn what is going to happen. Insurance companies that want to reduce the number of likely causes of death require physical exams of persons before deciding the price to sell an assured payoff for loss of life. Games usually play heavily on the drama to see which contestant does the best at achieving some desired end.

CHAPTER 12.0

Current and pressing questions are: When will the world run out of oil? And how long until world oil production fails to keep pace with demand? The world will not run out of petroleum in the foreseeable future but it became too valuable to burn decades ago when polymers were invented. Oil has continued to fuel progress but at a far greater cost than renewable fuels would have. Each improved polymer invention and the world's steady population growth translates to an increase in the opportunity cost of burning oil. We are now at the momentous transition from a buyer's market to a seller's market because world oil production is failing to keep up with demand. The world now has too many oil, gasoline, and diesel fuel users for the world's limited production capacity ... and the production capacity is tapering off because of depletion while the population continues to increase.

Civilization has decided to pay much attention to opportunistic control of scarcity to help the Haves derive greater profits from the Have-nots. Haves are encouraging the Have-nots to stay dependent upon finite supplies of fossil hydrocarbons and are ridiculing the notion that fossil supplies are running out.

Have-nots are becoming aware that humans have burned much of the easily extracted oil that Nature took 500 million years to accumulate. Have-nots in countries that export much of their oil notice the depletion and resent the fact that someone far away is getting most of the benefit. It is not surprising to find Have-nots that resent the energy-intensive lifestyles of industrialized countries.

America once had about as much petroleum as Saudi Arabia but now the U.S. has less than 2.5% of the world's remaining oil reserves. We burned coal, oil, and natural gas faster than any other country to influence world affairs and to support enormous consumerism. Much of this consumerism is based on keeping up with the Jones. In the 1890's the oil age was young and 11 million of the 12 million families in the U.S. earned less than $1,200 per year.

CONSPICUOUS CONSUMPTION:
Is spending conspicuously ... the end?

Imagine someone spending more for an evening party than most families could expect to earn in 308 years! "Conspicuous consumption" described what economist Thorstein Veblen found amazing regarding an 1899 ball that was hosted by Mrs. Cornelius Sherman Martin for about 900 privileged persons. This ball cost $370,000 or $7 million in today's inflated dollars. Dr. Veblen abhorred waste but realized that such excesses by the wealthy served a purpose. It stirred the lower class to find ways to become wealthy.

Veblen observed that in the emerging energy-rich American economy, the lower class did not want to overthrow the upper class. Lower class persons were striving to join the wealthy. Lower class persons were planning to join and then depose the upper class by outspending them. This dream had sound basis because America had plenty of land for farming, extensive timber resources, and mining opportunities for a vast assortment of minerals including coal, oil, and natural gas that could be burned to mechanize virtually any pursuit of quick wealth. After a century of exploitation by the legions of new wealth achievers and conspicuous consumers this version of the American Dream is on the block.

Continued conspicuous consumption of the present kind by those that seek status by consumerism will increasingly encounter the problem of Peak Oil as demands for oil exceeds the ability to produce it. Peak Oil shifted the economy from the buyers' market that launched a century of consumerism to a sellers' market. Energy companies that have been breaking records for income production per employee will be able to take greater profits on oil production from remaining oil reserves. Transition to a sellers' market will cause increased strife between OPEC and Have-nots unless and until the oil Have-nots decide to adopt renewable resources and develop a sustainable economy.

Recent declarations that we are in the computer information age conveniently forget the fact that this age, like its continuing predecessor the automobile age, was designed to depend upon fossil fuels. The Internet computer explosion now demands a sizeable portion of electricity production. At the point that one billion PCs are on the World Wide Web, more electricity will be required than the present U.S. electricity production. This burgeoning market for PC status, communication, and entertainment has accelerated the depletion of fossil and nuclear fuels, but will the Web provide increased communication to help develop worldwide awareness of the necessity to accelerate development of large-scale renewable energy projects?

Where will this take us before new inventions reshape our lives? So far Civilization has been inclined to embrace inventions that solve immediate problems or provide other forms of prompt gratification. Inventions that solve problems that are too far in the future are much harder to implement. And why not conspicuously consume the remaining fossil reserves and hope that someone will invent *(but only in the nick of time)* some technology that will enable at least some of our kind to travel thousands of light years to populate another planet? Won't we need this anyway to escape from some catastrophic end of the Earth? Science fiction glamorizes this approach but the distance to another planet with Earth's gravity, atmosphere, water resources, rocky soil, etc., is probably many thousands if not millions of light years away.

Who will go to represent the 7 billion left behind knowing that their frozen body, reproductive cells, or dust will be required to travel for millions or perhaps billions of years to somewhere far beyond present comprehension? With each improvement in telescopes, the perceived Universe expands. The Universe as presently perceived has multitudes of galaxies that are detected at distances up to 15 billion light years from our Sun.

HOW WILL THE EARTH END?
Is the end of life on Earth … the end?

The Earth as we know it could be around for several more billion years but life on Earth and the pursuit of human happiness faces challenges. It is probable that life on Earth will be threatened by human errors and/or by collision with something from the dark reaches of space that is traveling at a sufficiently high relative velocity to cause overwhelmingly catastrophic events on Earth. Planetary physicists believe the Moon formed from matter that was torn from the Earth by a collision, or very near collision, eons ago.

12.2

Collision or near collision with an object large enough to tear away the mass that became the Moon would cause sudden heating of the oceans and atmosphere. It would kill most forms of life on Earth. Most living things would be denatured by the collision blast and resulting global atmospheric heating. Life as we know it depends upon bio-diversity. Even if microbes that lay dormant in deep subterranean aquifers or ocean ooze survived the surface catastrophe, they would lose their lifeline of complementary forms of life. Diligent planning, international cooperation, and timely preparation can prevent this catastrophe.

We will be well advised to build robotized rocket ships for intercepting, diverting, and/or mining any dangerous comet or asteroid that threatens to collide with Earth. Unmanned space ships could be built by extending the success of the Apollo Program and the Saturn Technologies. Based on the existence of our Moon and the collision craters evident on the Moon and the Earth we can be sure that there are serious collision threats to our well-being. Fortunately one of Americas' conspicuous consumptions was to keep up with the former USSR in the

Space Race. Both contenders in the Space Race excelled and Earth has proven technology that can be improved for handling the threat of collision with an object from space. We have the option to make adequate preparations and survive until a more overwhelming catastrophe confronts Earth.

Pondering the mystery of how our Earth will end begs the question of our planet's beginning. In turn, this requires asking the same question about our Galaxy which is estimated to be about 16 billion years old. During these 16 billion years since the beginning of time, some 10^{21} stars have formed. (1,000,000,000,000,-000,000,000 stars!) But how can the "end" be guessed without knowing much more about the beginning of the Universe or the beginning of time, or where our small solar system is among the multitude of stars in the Milky Way Galaxy and where it is in the Universe? A consensus has emerged that theorizes the beginning of the material universe as an explosion of subatomic particles that produced protons and electrons to form hydrogen. When sufficiently large amounts of hydrogen accumulated, stars were formed. In the pandemonium of star physics, stupendous gravitational pressures and extreme temperatures, one hundred twelve (112) other elements were produced by rearrangement of the particles.

The Milky Way Galaxy of some 400 billion stars rotates about a Galactic center. Our sun along with nine planets in gravitational tow are in the outer region of the Milky Way's multitude of rotating stars that form a cosmic cloud about 3,000 light years thick in the center and 100,000 light years in diameter. Our Sun is about 28,000 light years from the Galactic center of rotation. Each rotation of our Solar System around the Galactic center takes about 200 million years. Fossil deposits were largely established from about 600 million years ago to about 60 million years ago.

FIGURE 12.1

The Milky Way Galaxy

Earth physics is far too cool and has far too little pressure to produce the elements that are found here. Observations of far away stars that blew up led to the realization that the amazing array of elements found on Earth are parts of the blast debris from stars that existed long ago. Speculation about the age of our Sun generally indicates that our midlife sun should remain stable for 4 or 5 billion years before it matures into another phase of fusion and destroys life on Earth. But this ominous "end" by being engulfed in an inferno of nuclear fusion is ironically offset by predictions of a more ultimate end of the Universe as death by cooling.

Observations seem to hold that the entire Universe, at least all that we can perceive is rapidly expanding. Expansion increases the distance between the stars and all other material components of the Universe. Such expansion reduces the gravitational force available for collection of matter. As existing stars mature and either explode or burn out as fusible matter is depleted, the Universe cools but continues to expand. The end of the Universe seems to be endless expansion and ultimate cooling towards absolute zero. So, the Norse might have been right by perceiving Hell as dreadful cold beneath an empty sky.

With this ominous future of adverse predictable circumstances in store for our descendants, what is the best hope for honorable, meaningful, existence for those that are presently alive? Surely we are best served by making the most of our circumstances.

We must find ways to extend prosperous human representation into the future as long as it is possible. It is the ultimate instinctive purpose that distinguishes all forms of life from the non-living. Every living thing is designed to extend life into the future, and humans have been winning the race to use more of the Earth's resources to do so. It seems to be a common belief that humans are more capable, more deserving, and more aware of a purpose for doing so than other life forms. But what is the purpose? Does Civilization have a purpose?

OUR GRAND PURPOSE:

12.3 The arrow of progress points the way. Progress in human's search for the good life took a decidedly beneficial turn when the fossil reserves were discovered. With many orders of magnitude more disposable energy per person, progress was rapidly achieved. Disposable energy delivered by pipelines and electricity grids enabled individuals to spend more time developing specialized talents and capabilities instead of collecting wood or shoveling coal. A relatively small scattering of industrious farmers with fossil-fueled equipment could feed the cities. Manufacturers were able to quickly provide building products, clothing, appliances, and vehicles for the burgeoning population.

Our Grand Purpose is to achieve sustainable prosperity and this purpose will be served by continuing to increase the amount of disposable energy per person. We will accomplish this by harnessing the enormous potential of solar energy and its derivatives such as wind, wave, falling water, and biomass.

History correlates the rate of technology development with the amount of disposable energy available along with the higher standard of living that it offers. We will need many new inventions to facilitate the achievement of sustainable prosperity. The more hydrogen, carbon, and other renewable resources that we can develop, the more likely we will sustain the pace of technology development required to support global prosperity.

Our place in the Universe is Earth. We must cherish and make the most of our circumstances. We already have technology to do much of what is needed. But we need to apply key technologies in large-scale production of renewable energy. We know that relatively small amounts of land or sea can supply more energy than required to achieve comfortable, meaningful, and inspiring life styles for Earth's existing population.

The sooner we convert the engines of the transportation sector to hydrogen operation and shift to distributed cogeneration of heat and electricity using hydrogen, the better it will be for present and future generations. If we work on it soon enough and diligently enough, the Solar Hydrogen Civilization will prevent enormous hardship and conflict as it inspires achievement of our Grand Purpose of sustainable prosperity.

Our quest must continue without tiring until we achieve the Solar Hydrogen Civilization. This new horizon of inventions for Civilization will provide endless improvements in living standards, health, security, and enjoyment of life.

DEVELOPING THE SOLAR HYDROGEN INFRASTRUCTURE:

WITHIN A FEW YEARS: Provide mass production of equipment and packaged plants for converting methane from agricultural wastes, sewage, and garbage into hydrogen and carbon. Provide equipment and packaged plants for adding methane to the natural gas pipeline distribution system. Develop electronic energy contract hardware and software. Develop hydrogen cogeneration systems. Develop Clean-Air Lotteries in planned communities. Convert vehicle fleets to operation on hydrogen.

12.4

Develop Solar Hydrogen Corridors:

Belts of highways, railroads, pipelines, and the electricity grid, extend from coast to coast and across America. A hydrogen corridor provides renewable methane and/or hydrogen for distribution through existing natural gas pipelines. Solar Hydrogen Industrial Parks and Solar Hydrogen Service Stations convert methane to hydrogen and carbon products. Homes and businesses utilize off-peak electricity to provide pressurized supplies of hydrogen and oxygen by electrolysis. An early Solar Hydrogen Corridor could extend from Los Angeles to Washington D.C. to be joined with solar corridors along Highway 5 along the Pacific Coast and Highway 95 along the Atlantic coast. Middle America could be joined by corridors along Highway 25, I-40 and the old Route 66.

See Appendix

WITHIN A DECADE: Develop storage interfaces for methane and/or hydrogen in depleted natural gas and oil reservoirs. Specify Cogeneration and/or Total Energy Systems for government buildings. Order hydrogen powered vehicles for government vehicle fleets. Large-scale Renewable Resources Parks will convert Solar, Wind, Wave, and Biomass Resources into renewable energy and materials.

CONSUMER DRIVEN CHANGES:

12.5 In an efficient economy, consumers have the last word regarding product selections and consequent exploitation of resources. The world needs conspicuous consumerism of renewable goods and services that overcome our impending economic crisis of dependence upon burning fossil fuels. Near future products and services that will be provided by new ventures that are launched by the International Renewable Resources Institute include the following:

HYDROGEN SERVICE STATIONS:

12.6 Service stations will receive methane that is delivered by natural gas pipelines and convert it to hydrogen and carbon. Equation 12.1 summarizes the endothermic process:

Equation 12.1

$$CH_4 + HEAT \rightarrow 2H_2 + CARBON$$

Heat for the endothermic process of dissociating methane into carbon and hydrogen may be supplied by electric induction or resistive heating, concentrated solar energy, or by combustion of methane, hydrogen, or carbon. Energy requirements are much less than required for dissociation of water. About 285 kJ/mol H_2 must be provided to dissociate water compared to only 37.4 KJ/mol H_2 for dissociation of methane.

In the development phase of the renewable infrastructure, dissociation of methane can be accomplished by burning a relatively small portion of the incoming methane to produce

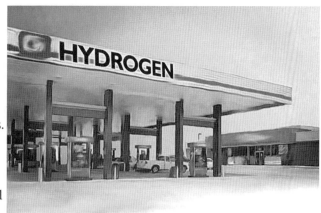

the required energy for production of hydrogen and carbon. Even so, this causes only about 10% of the carbon dioxide that is released by conventional burning of natural gas. About 90% of the carbon can be sequestered for production of durable goods.

Combustion of a relatively small portion of the methane releases enough heat to dissociate the remaining methane. Heat is needed to drive the process so there will be little heat loss allowed from the well-insulated carbon sequestration units. As Renewable Resources Parks develop, carbon dioxide produced by combustion of methane will be eliminated as dissociation energy is supplied by renewable electricity or direct application of concentrated solar energy.

FIGURE 12.3

Packaged carbon sequestration units of various sizes will be delivered to service stations, farms, and factories. These packaged carbon sequestration units will be transported by truck, connected up to a natural gas line, and quickly start production of hydrogen and carbon. Hydrogen in various grades and "recipes" can be provided. The highest purity hydrogen will be provided for low-temperature fuel cells. Lower purity hydrogen will be used to fuel internal combustion engines. Hy-Boost mixtures of 5 to 95% hydrogen with methane will fuel internal combustion engines that operate much more efficiently with un-throttled air entry and stratified-charge combustion.

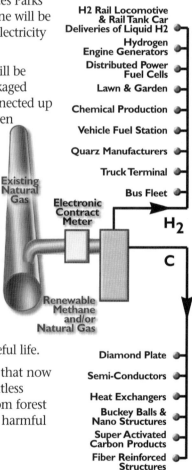

Carbon sequestered by dissociation of methane can be delivered to industrial parks for making a multitude of new products including high capacity hydrogen storage tanks, sporting goods, architectural components, machine tools, marine vessels, and other vehicle components that reduce curb weight and extend useful life.

This will provide market pull for renewable methane that now goes to waste as biomass rots into the atmosphere on countless farms, at city sewage and garbage disposal facilities, and from forest wastes that pose the double dangers of accidental fires and harmful greenhouse gas emissions.

SEQUESTRATED CARBON:
for lighter, stronger, safer, and more durable products.

New ventures at industrial parks that process renewable methane will make sporting goods, transportation components, and process equipment. Utilization of stronger and lighter carbon reinforced materials to reduce curb weight of vehicles provides many benefits. Hydrogen produced by dissociation of renewable methane and/or fossil natural gas will be used to produce electricity and for area vehicles.

12.7

Fuel economy in stop-and-start driving benefits from reductions in curb weight because less energy is lost overcoming inertia to achieve suitable acceleration to desired speeds. In many vehicle types and driving conditions, each 10 percent reduction in curb weight provides a 7 percent improvement in fuel economy. Performance is also improved with reduced curb weight. Vehicles accelerate more rapidly, turn better and stop more quickly for improved drivability and safety.

One of the important benefits of carbon will be fiber reinforcement for greater crash safety.

Solar collection systems, wind machines, in-stream hydroelectric generators, wave machines, homes and other structures will also be improved by new carbon-based products that are stronger, more fatigue resistant, and far more durable in corrosive environments.

Watch for multitudes of new products that are based on large-scale and nano-scale applications of superactivated carbons, diamond plating, and carbon-crystal reinforced developments.

Electronic contracts facilitate distributed energy:

Two types of electronic meters will help launch the Solar Hydrogen Civilization. One type measures the amount of energy that is added to or taken from natural gas pipelines. The other meters electricity that is added to or taken from electricity grids.

Farmers and others that produce biomass wastes such as animal manure and crop wastes can enter into electronic contracts with industrial park operators and city dwellers. Farmers will operate anaerobic processors to produce pipeline quality methane, which is electronically metered into the natural gas pipelines.

An electronic contract with a customer who is served by the pipeline allows a measured amount of methane to be taken from the pipeline which is less than the amount added by the farmer. The difference between the amount of methane that the farmer adds and the customer receives as recorded by electronic accounting is the amount of methane paid to the pipeline operator. This payment of methane covers the costs of maintaining and operating the pipeline along with an allowance for extending the pipeline to new customers.

Customers can set electronic bids to buy methane for a certain time span at a certain contract price. Methane providers can determine if they want to take the contract and incur the methane cost that is bid by the pipeline operator for delivery. Shorter distance deliveries will be favored but the amount of methane collected by the pipeline company can be sold to other customers or

FIGURE 12.4

Biomass from farms (cattle, poultry and swine as well as crop waste) is processed to produce methane and sent through the existing natural gas pipeline to an industrial park which dissociates the methane into hydrogen, which is used in fuel cells and engines and carbon for producing durable goods.

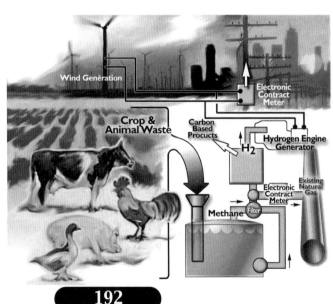

traded to other pipeline companies. Automatic monthly statements will be tallied to show the details of the transactions and to establish payment records.

Hydrogen, co-produced with carbon at industrial parks will fuel hydrogen engines that clean the air. It will also power fuel cells along with countless other applications for more profits.

Another type of electronic contract meter will measure power that is added by fuel cells and hydrogen engine-gensets to the electric grid. Electronic contracts will provide for deliveries of electricity to customers on the grid. An amount of electricity proportional to the distance between producers and customers will be paid to the electric grid operator for maintaining and managing the grid along with extending the grid to new customers.

Electronic contracts for methane and electricity deliveries will provide incentives for farmers, city managers, and entrepreneurs to develop important income streams from solar, wind, hydro, and biomass waste resources. Doing so will expedite the Solar Hydrogen Civilization and provide much greater energy security along with anti-inflationary benefits.

FIGURE 12.5

Using the TESI engine in conjunction with a heat exchanger, adapted home appliances can utilize heat transferred from clean exhaust air.

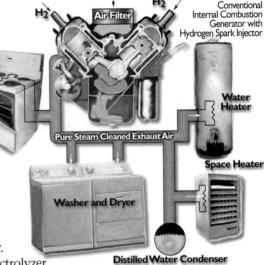

Total Energy System Innovation (TESI) and TESO Systems:

Long-life engine-generators will produce electricity for your new hydrogen service center. When surplus-generating capacity exists, an electrolyzer produces high purity hydrogen for the most requiring industrial, pharmaceutical, semiconductor, and fuel cell applications. A thermochemical regenerator utilizes heat cascaded from the engine to produce Hy-Boost fuel for your TESI engine and for your customer's engines or more robust fuel cells. TESI operates efficiently on hydrogen, biomass sources methane, natural gas, propane or butane.

TESO also utilizes a long life TESI engine-generator and is equipped with a thermal dissociation unit to convert biomass feed stocks such as crop and forest wastes into Hy-Boost fuel for its engine and customer's engines.

EC-TES: ELECTRO CHEMICAL TOTAL ENERGY SYSTEM

An electrolyzer cabinet about 24" wide, by 24" deep by 48" high will be connected to potable water and electricity to produce enough hydrogen for two vehicles and the home or small business where it is located.

12.8

The reversible electrolyzer receives electricity from locally produced solar-electric, wind, wave, or falling-water generators or by electronic-contract arrangements and produces hydrogen and oxygen from water. At times that electricity for the home is needed, a portion of the reversible electrolyzer will be operated as a fuel cell to convert stored hydrogen and oxygen into electricity and water.

Pressurized supplies of oxygen will be produced for industrial and medical applications. Hydrogen-Oxygen torches will replace Oxygen-Acetylene welding and cutting torches. Medical supplies of oxygen will be produced on site at a much higher overall energy utilization efficiency than by conventional fossil-dependent methods.

RETROFIT DIESEL ENGINES FOR HYDROGEN OPERATION:

12.9 Truck transportation to and from specific addresses on city streets provides significant convenience and along with farming, rail and marine transportation accounts for most of the diesel fuel demand. Diesel engines are preferred for powering large engines because of superior fuel economy compared to homogeneous charge gasoline engines.

However, numerous recent studies link particulates from Diesel engines and coal-fired power plants to asthma, pneumonia, cancer, and heart diseases. Deaths due to asthma increased 118% from 1980 to 1993 and this increase is correlated to increased childhood exposure to particulate contaminants in the air. Some 127 million persons in the U.S. live in areas that do not meet federal clean-air standards. In 1999, the World Health Organization estimated that more than 70,000 deaths were due to diseases caused by breathing air contaminated particles with aerodynamic diameters less than 2.5 microns.

Diesel-fueled engines produce over 100 times more such exhaust particles than gasoline engines per horsepower-hour of use. Diesel engines source about 26% of the total hazardous particulate pollution from all fuel sources and 66% of the particle pollution from

FIGURE 12.8 *Hydrogen is generated for the TESI engine through the use of Bio-Waste (manure, forest waste and garbage) and the Carbon Sequestration process. Excess Hydrogen, Methane and Electricity is then sold to customers and transmission utilities.*

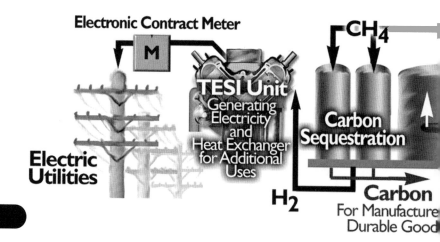

motor vehicles. Diesels emit about 20% of the total nitrogen oxides (NOx) from all outdoor sources and 26% of the on-road sources. Persons driving equipment powered by Diesel engines, working in places that service such engines, and persons who live near distribution centers that have frequent Diesel-truck activity may be exposed to air contamination levels of Diesel particulate emissions of more than 3.6 micrograms per cubic meter. EPA estimates indicate that more than 350 million people could develop cancer during a lifetime of exposure to air contaminated with 2.2 milligrams per cubic meter of Diesel particles.

Some 500,000 diesel engines in trucks and farm equipment can be retrofitted to utilize hydrogen and/or Hy-Boost fuels. Conversion to hydrogen and/or Hy-Boost fuels will enable existing Diesel engines to clean the air while producing equivalent or greater power when needed.

FIGURE 12.8

Modular unit with connection to natural gas line delivers hydrogen to a fuel dispenser that loads tractor-trailer trucks and rail cars.

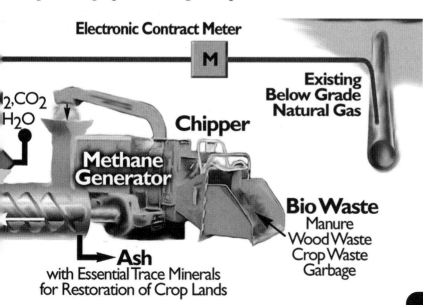

RAIL LOCOMOTIVES CLEAN THE AIR
AND DELIVER RENEWABLE FUELS:

12.10 Over 20,000 rail locomotives will be converted to utilize renewable hydrogen and/or Hy-boost fuels. Conversion of this important fleet will enable annual fuel savings of more than two billion dollars and reduce atmospheric pollution by at least 100 billion pounds of carbon dioxide.

New technologies allow hydrogen and Hy-Boost fuels to eliminate particulates, carcinogens, and emit far less oxides of nitrogen when utilized to convert Diesel engines to clean operation. Diesel engines converted to operation on hydrogen and/or Hy-Boost fuels can produce more power; last longer, and actually clean the air. Proven technologies provide for direct-injection and stratified-charge combustion to improve on existing fuel economy advantages of Diesel engines.

Rail delivery of commodity supplies of hydrogen and Hy-Boost fuels will be facilitated by conversion of the Diesel locomotives now powering trains across the continent. Converting rail locomotives to hydrogen operation will reduce growing requirements for importing oil by more than five billion gallons of fuel each year for use by rail locomotives. Much larger reductions in imported oil and emissions will be achieved by rail deliveries of hydrogen for trucking applications in locations that are not served by pipelines.

HYDROGEN POWERED TRAVEL THROUGH
AIR AND SPACE:

12.11 Hydrogen-powered spacecraft hold the record for greatest non-stop distance of travel. NASA's visits to the Moon and back by hydrogen-oxygen rockets far exceed all previous travel distances on Earth. Space travel requires the lowest possible curb weight. Hydrogen was chosen for these rockets because it yields about 3-times more combustion energy per mass unit than hydrocarbon fuels, it remains liquid at the very cold temperature of –421°F (39°R), and it is readily utilized for both the propulsion rockets and the life-support fuel cells that provide electricity and potable water.

Air breathing airplanes will accomplish similar benefits by utilizing hydrogen in place of hydrocarbons such as jet fuel, kerosene, or gasoline. Aircraft using piston engines, gas turbines, and various advanced-design engines will be fueled with hydrogen which will enable lower take-off weights for longer range flights with greater payloads. Noise generation will be less from aircraft that use hydrogen because lifting surfaces can be smaller and more efficient with lighter fuel loads and the power required per payload can be greatly reduced.

Hydrogen-powered aircraft will clean the air that passes through their engines.

MOTOR VEHICLES CAN PRODUCE
INCOME WHILE YOU SLEEP:

Most of the world's multitude of vehicles do one thing... occasionally! They take one or more humans from point A to point B. The rest of the time these expensive gasoline-guzzling vehicles just sit and wait. A vehicle that accumulates 100,000 miles in 5 years actually sits still far more than it is driven. At an average driving speed of 27 miles per hour the vehicle would sit still about 40,000 hours between excursions that accumulate to about 3,700 hours in motion. Vehicles are often parked all day in parking lots waiting for someone to go home. All night they wait for someone to wake up and go to work or shopping. But most of the time they do nothing, but take up space while they depreciate and stay ready to wear out the tires and convert the costly gasoline that you buy into carbon dioxide and other pollutants.

New inventions offer to transform your car into a much better friend, one that earns money for you and relieves geopolitical strife. Imagine your vehicle coming home from a "tune up" with new capabilities that enable it to produce the electricity and heat for your home. A system of inventions can be installed during a future tune up to enable an overall energy utilization efficiency that is two times higher than the electric company that only sends about "X" amount of electricity for every "2X" amount of energy that they throw away on your behalf.

At the tune up, your car will be provided with SparkInjector Plugs that replace your ordinary spark plugs. SparkInjectors are named for their function, spark ignition and fuel injection. This enables your car to run just like it did from the factory on gasoline ignited by SparkInjector plugs or you can power your engine with much cleaner and less expensive fuels that are injected directly into the combustion chambers. Direct injection provides much greater fuel efficiency and enables your engine to use less expensive hydrogen or mixtures of hydrogen and landfill methane or natural gas. Operation in the SparkInjection mode will enable your engine to last much longer and do much more for you than it did when you bought it.

At night or in future parking lots your car can be earning money. An electronic contract plug-in at your parking spot will connect your car's SparkInjectors to a natural gas line and your car's electrical system to the local electric grid. The same plug-in will connect your car's cooling system to a heat exchanger that first receives heat from your engine coolant and then exhaust heat to provide hot water for heating the factory, office, greenhouse, or your home depending upon how many places you decide to have your car earn income or save money for you. You can also select an electronic meter that will keep track of your earnings after automatically paying your fuel bills.

After you are comfortable with the new arrangements for having your car earn income while you help double America's energy utilization efficiency, you may decide to install an electrolyzer in your home or use one near your parking lot to supply some of the fuel your car requires between parking places. Think of it, you will be using cheaper fuel that is increasingly produced by farmers and city managers who send renewable methane into the natural gas lines to make ultimately clean hydrogen to power your car on the roads and highways. Emission tests show that car engines converted to SparkInjector operation with hydrogen can actually clean the air and produce exhaust gases (air and dihydrogen monoxide -- also known as steam) that are cleaner than the air that enters the engine. Existing engines can last longer, produce more power, and steam clean the air...and produce income for you as they provide market pull to help America achieve sustainable energy independence.

You will be doing more for Civilization than all the marching armies since King Ramses II and more good than all the economists of the Industrial Revolution who advocated the economy of fossil depletion as a desirable pursuit.

C O N C L U S I O N :

Until the stars no longer support human endeavors, inventions will provide endless progress by the Solar Hydrogen Civilization.

Glossary

A

Abundant energy: Energy adequate to meet present and future needs. Solar energy is far more abundant than any other energy on Earth. Solar energy delivered to the Earth each day is equivalent to 2.5 trillion barrels of oil (14.7 x 1018 BTU) or about 1.4 times the energy in all the 1.75 trillion barrels of oil that ever existed (equivalent to 10.15 x 101^8 BTU). Solar energy is widely distributed in winds and waves, falling water, sun-baked deserts, and biomass resources that can be harnessed virtually without pollution and without depletion.

Aerodynamic control: Performance of an airfoil (geometry, modulus, structure, and surface) such as a propeller design and its response to varying velocities of airflow. Such as an overspeed-protection feature for a wind generator.

Ampere: Unit of electrical current measurement.

Alternator: An electric generator for producing alternating current.

Alternating current: Current that cyclically flows in one direction and then in the other.

Ampere-hour: A quantity of electricity equivalent to a current of one ampere for one hour.

Anemometer: An instrument for measuring the velocity of the wind.

Armature: A drum-like cylinder composed of a multitude of thin sheets of steel and wrapped with wires that carry current to create a magnetic field or pass through a magnetic field. An armature produces electricity in a generator and produces torque in a motor.

Avogadro's Law: At the same temperature and pressure, equal volumes of all gases have the same number of molecules.

Avogadro's Number: The number (6.022 x 1023) of atoms, molecules, or particles found in exactly 1 mole of a substance.

B

Barrier to entry: Any technical, institutional, governmental, or economic restriction that delays or inhibits product entry into a market. Primary barriers include resource control, monopolies, patents and copyrights, government restrictions, and availability of capital. Monopolies, oligopolies, monopsonies and oligopsonies usually produce or are established by one or more barriers to market entry.

Battery: A device that converts a chemical potential between two electrodes separated by an electrolyte into an electropotential to produce electron flow in an external circuit.

Battery bank: A group of batteries connected in series and/or parallel to store electrical energy for later use.

Blade tip spoiler: An overspeed protection device: an airfoil attached to the tip of a propeller that is used to increase drag in high velocity wind and prevent overspeed of propeller.

Brushes: Usually rectangular shaped carbon that contacts the commutator or slip-rings to transmit electricity into or out of an armature.

Brushless exciter: An AC generator with a rotating armature and a stationary field mounted on the main shaft of a generator.

Biomass: Direct or derivative product of photosynthesis. Organic substances are characteristically made of carbon, hydrogen, oxygen, and lesser amounts of other elements such as nitrogen, sulfur, calcium, phosphorous, potassium, iron and trace elements.

Centrifugal blade pitching: An overspeed protection system: a spring and weight mechanism attached to propellers for changing the angle, or pitch of the blade in higher than rated velocity winds.

C

Charge: To restore the active materials (in storage battery) by the passage of a direct current in the opposite direction of discharge. To add kinetic energy to an energy-storage flywheel. To add hydrogen and/or oxygen to an energy-storage system. To increase the pressure of an accumulator. Addition of electrons to a capacitor.

Closed circuit: Current flowing in a complete path from the power source, through a conductor to the load and back to the source.

Combustion: Exothermic chain reaction by a fuel and oxidant.

Commutator: A ring of the copper wire endings on the surface of an armature. When the armature turns, the commutator is contacted by brushes that slide over it and transmit the generated electricity to external applications.

Compression ignition: Fuel ignition by hot air or other oxidant that has been heated as a result of compression. Diesel engines ignite injected diesel fuel in air that has been heated by rapid compression.

Conductor: Any material that permits the free flow of electrons or other electrical charges.

Constant frequency: A regular and consistent number of alternations per second of a current (U.S. standard is 60/minute, other countries may be 50/minute).

Controls: A mechanism used to regulate the operation of a device such as a generator, i.e., (voltage, frequency, shutdown, utility connect and disconnect, AC or DC mode).

Coriolis force: An apparent force that as a result of the Earth's rotation from west to east that imparts force on moving objects (as projectiles or air currents). Winds tend to rotate counterclockwise about low-pressure centers in the Northern Hemisphere and clockwise about low-pressure centers in the Southern Hemisphere.

D

Darrieus: A vertical axis wind turbine system invented by G. M. Darrieus of France in 1925 (sometimes referred to as the "eggbeater" because of the rotor shape and appearance).

Diode: A rectifier device that blocks current in one direction and passes current in the opposite direction.

Direct current: Current that flows in one direction only.

Discharge: Dissipation of a charge, such as the conversion of the chemical energy of a battery into electrical energy. Depressurization of a pneumatic accumulator. Conversion of potential energy of a battery or charged capacitor into electrical current that does work or generates heat.

Downwind: Vector in the direction of the wind travel.

Dynamic braking: The machine acts as an generator, converting the kinetic energy to stored energy or dissipating kinetic energy of rotation either as heat in a braking resistor or in a direct short circuit on the generator.

E

Economic goals: Stability, sustainability, efficiency, equity, full employment, peace, and satisfaction.

Economic inflation: Price escalation due to inability or unwillingness of suppliers to meet market demand. Energy intensive goods and services that require fossil or radioactive energy must ultimately be subject to price controls, rationing, or free market inflation because of depletion of finite fossil and radioactive energy sources. It is mean-minded to curb inflation by creating unemployment thus limiting the ability of buyers to seek limited supplies of desired goods. It is far better to welcome and facilitate virtually full employment to utilize sustainable energy and material resources to provide goods and services demanded. Employment to increase production of goods and services from renewable resources sustains economic growth. Every community can prosper in the "More-Tomorrow" Revolution.

Economy: Production and distribution of wealth. An interaction of producers and users in which a market structure can be identified and characterized by one or more of four basic operations. Perfect competition, monopoly, oligopoly, and monopolistic competition are these basic market situations.

Efficiency: Measure of desired result compared to the resources expended to secure the result. An total energy system with an engine that burns renewable hydrogen to provide 30 units of work and 70 units of heat that is cascaded through heat exchangers to provide 50 units of heat for beneficial results such as food cooking and preservation, crop drying, water heating, and maintenance of an anaerobic digester could be said to provide an energy utilization efficiency of 80%. Additional satisfaction could be attributed to cleaning the air that passes through the combustion chambers and beneficial utilization of potable water that is condensed from the lowest temperature heat exchanger. The efficiency of an economy is found by measurement of the satisfaction of needs and wants by selection of options that produce the most benefit with the least expenditure of resources. An economy that spends enormous amounts to find, burn, and deplete in 500 years the fossil substances that took Nature 500 million years to accumulate is not as efficient as an economy that satisfies needs and wants by utilizing sustainable energy sources.

Electrically actuated tail control: A small servo-motor control folds the tail horizontally to turn the generator out of the wind.

Electricity: The accumulation or flow of charged particles such as electrons through a conductor.

Electrodes: Conductors used to establish electrical contact with a nonmetallic part of a circuit.

Electrolyte: A nonmetallic electric conductor in which current is carried by the movement of ions.

Electrolysis: Production of chemical changes by passage of an electric current through an electrolyte.

Electromagnet: A magnet that has been given increased magnetic properties by passing a current through insulated wire to produce a magnetic field.

Electron: Invisible, negatively charged particle that orbits the nucleus of an atom. Smallest negatively charged particle of electric current.

Enthalpy (H): Measure of the sensible heat in a process. The enthalpy of formation of elements is arbitrarily set at zero and all compounds have an enthalpy of formation that is measured from this reference.

Entropy (S): An intrinsic property which for systems at internal equilibrium that undergo a reversible processes is equivalent to: $dS' = dQrev /T$ where S' is the total entropy of the system, Qrev is the heat transfer, and T is the absolute temperature. Entropy is the measure of disorder of a system and is believed to increase if a spontaneous process occurs. Absolute entropy is the increase in disorder from zero of a perfectly ordered crystal at zero degrees Kelvin temperature to the condition at a given temperature above absolute zero.

Equalization charge: A charge to batteries that equalizes the charge on all cells and is capable of restoring 100% readiness.

Feathering: An overspeed protection that changes the angle or pitch of the blades to slow them in high velocity winds.

F

Frequency: the number of complete alternations per second of an alternating current.

Force field: Force imagined as lines of force which tend to pull or push one object or electron from another because of the alignment of their atomic electrons.

Glossary

G

Fuel cell: A special battery that is capable of being recharged with fuel and oxidant and can provide continuous electrical output so long as it is supplied with fuel and oxidant.

Generation components: Sub-elements of a wind system that actually contribute to producing electric power, including the generator or alternator, tower, rotor, and controls.

Generator: A device that produces electricity by repeatedly passing a conductor (wire) or series of conductors (wires) through the magnetic field of one or more magnets.

Generator field: The magnetic field of a generator.

Governor: An attachment to a machine for automatic control or limitation of speed.

Greenhouse gases: Gas species that absorb radiation from the Earth and/or from the Sun to cause energy to be added to the atmosphere.

Guyed tower: A tower that is supported or stabilized with one or more chains, wires, ropes or rods.

H

Hertz: Unit of frequency equal to one cycle per second.

Homogeneous charge engine: Engine that combusts homogeneously mixed fuel and air in the combustion chambers.

Horizontal axis: A wind generator that has provides an axis around which the turbine rotor revolves, parallel to the earth's surface.

Hydraulic: Operated by water or other liquid flowing through a pipe, channel or orifice.

Hydraulic blade pitch: Control of blade pitch or angle change operated through the use of hydraulics.

Hydro-electric: Production of electricity by waterpower.

I

Induction generator: An inductor stores energy in the form of a magnetic field. A conductive loop or coil stores energy in the magnetic field it produces. Inductance is proportional to the number of turns, the radius of curvature, the current, and the type of material around the current. An induction machine either squirrel cage type or wound rotor type that is forcibly driven above its synchronous speed at which point it acts as a generator of electrical power into whatever electric system it happens to be connected (assuming that electric system is energized and can supply the necessary excitation current to the induction machine.)

Inverter: A device for converting direct current to alternating current by mechanical or electronic means.

Insulator: A material that hinders or resists the free flow of electrons.

Isobars: An imaginary line or a line on a map or chart connecting or marking places on the surface of the earth where the barometric pressure is the same either at a given time or for a certain period.

Kilowatt : 1,000 watts.

Kilowatt-hour: A unit of work or energy equal to that expended by one kilowatt in one hour.

Lead-acid battery: An electrical storage device consisting of lead electrodes and sulfuric acid.

Load: An electrical draw.

Magnet: Any substance that attracts another substance through space by means of a magnetic force field.

Magnetic field: Envisioned as invisible lines of force which surround all magnets and which pass from the north pole to the south pole of the magnet.

Magnetism (magnetic force): The force effect of a magnet.

Mechanical air spoilers: Centrifugally actuated plates that move out at a predetermined wind speed to increase drag and effectively air brake the propeller.

Monopoly: A market situation that has a single seller of a unique product with no equally desirable alternatives. A selling-side domination of the market. A natural monopoly is a special monopoly that may lower the selling price as production efficiency increases with volume. One of the best attributes of high productivity tooling in the manufacture of goods from plentiful resources is the ability to produce very low cost items compared to hand-made and low rates of production from less-capable manufacturing systems. A single source natural monopoly results if two or more firms consolidate. After consolidation, at the sole discretion of the monopoly managers, may decide to increase or reduce the cost of each unit sold.

Monopsony: A market that has only one customer for a product. A buying-side compared to a selling-side monopoly.

Motors: Devices for converting electrical, hydraulic, or pneumatic potential energy into work.

Multi-bladed: Turbine consisting of two or more blades.

Natural magnet: Minerals, usually iron alloys or ceramics, such as magnetic lodestone, that are capable of exerting forces of attraction and repulsion on other ferro-magnetic substances.

Negatively charged: Having an excess of electrons.

Neutron: Invisible, neutrally charged particles found within the nucleus of an atom.

Ocean energy: Energy derived from waves, tides, ocean currents, or temperature gradient between one region and another such as surface and subsurface.

Ohm: Unit of resistance.

Oil Wars: Armed conflicts that consume vast amounts of petroleum to produce explosives, energy-intensive weapons,

J-K

L

M

N

O

food and clothing for military personnel, along with transportation equipment and fuel. World War I and all military conflicts since then have consumed enormous amounts of petroleum and coal resources and have been fought to control fossil resources along with other strategic purposes.

Oligopoly: A market structure dominated by a small number of large firms that sell identical or similar products in which there are significant barriers to new ventures that would enter the market.

Open circuit: Disrupted path of current.

Opportunity cost: The cost of sacrificing a choice when making a decision. Illustratively, deciding to burn one gallon of oil sacrifices environmental wealth and the opportunity to produce more than $35 worth of durable goods such as medical devices, electronic components, telephones, computers, fine clothing, and other desirable goods.

Overspeed protection: Protection designs that take effect at a predetermined speed for the purpose of protecting the generator from mechanical failure (flying apart) and/or to prevent an electrical supply overload.

Parallel circuit: Circuit that allows current to flow through more than one conductor path.

Peak Oil: Oil supply from a geological formation steadily increases until a plateau is reached. This is followed by a slow decline after the "peak," and then a steep decline. Crisis occurs when oil extraction can no longer meet world demand. Due to the finite nature of fossil fuels this marks the beginning of a permanent decline in oil output creating intense competition for short supplies. Competition and conflict grow in proportion to the decline in oil supplies and the increase in demand for oil. Peak oil imposes enormous economic and social disruptions and the more highly developed oil-dependent nations will suffer the greatest loss of productivity and economic security. Peak Oil is not about running out of oil; failure of production to meet demand happens much sooner and is exploited by profit taking by commodity traders and those in control of oil supplies.

Perfect competition: An ideal market structure that has a large number of small firms providing more or less identical products with freedom of entry into the market. Illustratively, a farmer's market without subsidies may approach perfect competition.

Petroleum: Oil and natural gas including methane, ethane, propane, butane, other paraffins and crude gasoline.

Permanent magnet: A magnet that retains its magnetism after removal of the magnetizing force.

Photon: Energy unit of light.

Positively charged: Having a shortage or deficiency of electrons.

Primary cell: Voltaic (battery) cells that are used up, when the chemical action between the electrodes and electrolyte ends.

Pressure gradient: The change in atmospheric pressure per unit of horizontal distance.

Propeller: A blade that interacts to be accelerated by relative motion with air, water, or some other fluid.

Proton: Positively charged particle normally found within the nucleus of an atom. Most hydrogen atoms have one proton as the nucleus.

Rectifiers: Devices that permit the flow of electrons in only one direction.

Resistance: Opposition to electron (current) flow.

Resistive load: An electrical load with the same time phase sequence of the current and the voltage. This type of load is usually made up of a straight current carrying conductor, electrolyte, or filament like the incandescent resistance element of a light bulb.

Ripple AC: Cyclical variation in rectified DC current.

Rotary inverter: Device having a combination motor generator with a governor that converts or inverts DC to AC.

Rotor(s): One or more blades that produce the torque such as the turbine driving a wind generator.

Sail wing: A propeller with cloth stretched over a metal wire frame that forms an airfoil section.

Savonius: A vertical axis rotor invented by S. I. Savonius that resembles two halves of a barrel in the letter "s" configuration.

Secondary cell: Battery that is capable of being recharged, i.e., where the elements can be restored to their original condition.

Self-supporting tower: A tower that receives its strength and rigidity from its structural design and firm anchoring. It requires no guying.

Series circuit: Current flow where only one conductive path is available.

Short circuit: Current path that bypasses the intended load.

Slip rings: Circular pieces of metal (usually a copper alloy) that connect a bundle of armature wires together to be contacted by the brushes, or which is used to transmit electricity if the armature wires are bound together by a single brush.

Sine wave: A cyclic waveform that represents periodic oscillations in which the amplitude of displacement at each point is proportional to the sine of the phase angle of the displacement.

Sine wave inverter: Usually a solid-state inverter designed with transistors or SCR's that produces a sine wave.

Solar cells: Devices that directly convert solar radiation into electricity without moving components.

Q-R

S

Solar hydrogen: Hydrogen produced by a suitable conversion of solar energy or a derivative of solar energy such as wind, waves, falling water or biomass resources.

Solid-state electronics: A very general term that indicates no moving parts and the use of transistors, photoelectric devices, silicon-controlled rectifiers, etc.

Solid-state inverter: An inverter designed with transistors or SCR's which has no moving parts.

Solidity ratio: Ratio of blade area to disc area.

Specific gravity: Ratio of the density of a substance to the density of a substance such as water or air.

Static electricity: Theoretically, electricity at rest or the collection of charge.

Stealing: Taking something in a way that causes hardship, anguish, and/or disappointment. Accounting the value of what is taken at the cost of taking instead of the replacement cost stimulates an inefficient economy and eventually causes hardship and disappointment.

Storage capacity: Number of amp hours a battery bank is capable of storing.

Storage component: Those sub-elements of a wind system which are used to store electricity and/or hydrogen.

SparkInjector©: Trade name of a proprietary device that replaces a spark plug or diesel fuel injector. SparkInjectors provide combined fuel-injection and spark-ignition functions to enable hydrogen and/or other spark-ignitable, inexpensive, clean fuels to replace fossil fuels. SparkInjectors facilitate unthrottled-air entry and stratified-charge operation of virtually any engine.

Stratified-charge engine: Engine that combusts fuel within surplus air.

Spark ignition: Application of a plasma to initiate combustion.

Synchronous inverter: A device that converts DC current to AC while regulating frequency to that of the power line.

T

Tail vane deflector: An overspeed protection device that provides spring loaded tails that fold in a horizontal plane to turn the machine out of the wind as the velocity increases past a limit.

Temporary magnet: Material which has magnetic properties when being brought near a magnetic field, but which loses these properties when the field is removed. "Soft iron."

Thermocouple: Device that produces electricity when two dissimilar materials are joined and heat is applied to the juncture. Current flow in a heated thermocouple junction is from the negatively charged material to the positively charged material.

Three-phase generator: Generator that produces AC power in three windings, which are 120 electrical degrees apart from each other.

Throttling loss: Loss of engine efficiency that is caused by restricting the air entry into the combustion chambers. Usually, throttling loss is due to efforts to provide homogeneous-charge operation and/or to intentionally reduce the power production from an engine.

Tip speed ratio: Ratio of the propeller tip speed to the wind speed.

Topography: Configuration of a surface, including its relief and the position of its natural and man-made features.

Torque: Force that produces or tends to produce rotation.

Tower shadow: Wind that has to pass the tower before it reaches the generator, i.e., the generator is in the wind shadow of the tower.

Transformers: Devices that increase or decrease current and/or voltage by electromagnetic induction.

Utility grid: Network of power lines, transformers, switches, and control systems.

U

Unthrottled: Intake of air or mixed air and fuel without restriction. Operation of an engine without intentionally producing or allowing an intake manifold vacuum.

Variable frequency: A varying number of complete alternations per second of an alternating current.

V

Vertical axis: A wind system that has its axis of rotation perpendicular to the earth and the wind velocity.

Voltage: (e) - Electromotive force: unit of electric pressure.

Voltage regulator: Device or design implementation that regulates the electromotive force (volts).

Voltaic cell: Another name for a battery. A primary cell named for its inventor Alessandro Volta.

Watt (w): Electrical power delivered by one amp at one volt.

W

Watt-hour: Unit of work or energy equivalent to the power of one watt operating for one hour.

Wealth: Total measure of assets. On Earth, the greatest asset is the environment that supports all forms of life and Civilization's economy.

Wealth expansion: Adding to the value, especially adding inventory and/or increasing the value of assets on Earth by utilization of solar energy or a derivative of solar energy to produce energy-intensive goods. Wealth depletion occurs as resources are sacrificed with less beneficial results than could have been achieved by utilization of renewable resources. Adverse changes to the environment by combustion of fossil substances illustrate wealth depletion as opposed to wealth expansion by utilization of solar hydrogen to produce goods and services.

Wheeling: The practice of sending energy as electricity or fuel through a suitable distribution network from a designated source to a designated receiver. Electronic contracts can be created by agreement between the designated source and the designated receiver and between the source and the operator of the distribution network. This arrangement may provide a larger amount of energy from the source than the amount received by the receiver and the operator of the distribution network is provided with the difference as payment for maintenance, operation, and expansion of the network to new customers.

Wind energy conversion system: A system of sub-components that converts wind energy to electrical and/or hydrogen energy. Tower, turbine, generator, energy conditioner, and storage system.

XYZ

SOLAR HYDROGEN CORRIDORS

The first coast-to-coast Solar Hydrogen Corridor could be from Los Angeles to Washington D.C. through Las Vegas, Albuquerque, Denver, Kansas City, and Indianapolis. This east-west corridor will link Solar Hydrogen corridors that cross the USA from San Diego to Seattle along Highway I-5; from Tucson through Phoenix and Salt Lake City to Helena; from San Antonio to Bismarck through Dallas, Oklahoma City, Kansas City; from Baton Rouge to Detroit through Memphis and Chicago; and along Highway I-95 from Bangor to Miami. Figure A-1 shows these early priority corridors.

Renewable Energy Parks will produce the hydrogen and/or methane and pipelines along interstate highways (HyWays) will transport these fuels to transportation, manufacturing, and electricity markets. Depleted natural gas reservoirs will store energy as needed to assure efficient response to seasonal demands.

FIGURE A-1

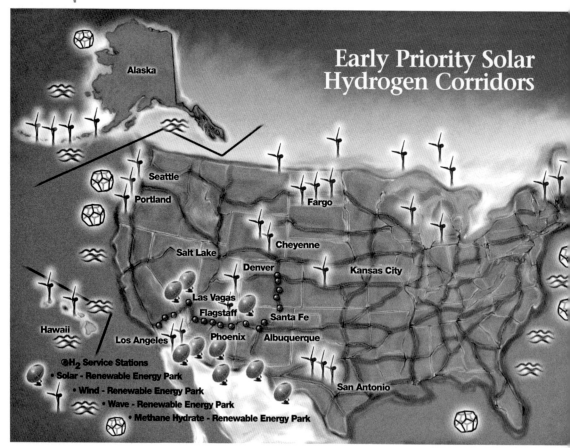

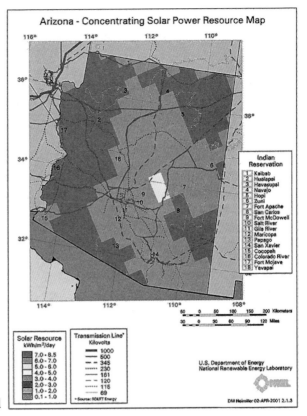

Arizona - Concentrating Solar Power Resource Map

In addition to hydrogen refueling stations in the towns and cities that are in these corridors, refueling stations will be spaced approximately every 100 miles along the highways of these corridors. Renewable Energy Parks will produce hydrogen and/or methane from solar, wind, wave, and biomass resources. Economic development will be supported as renewable resources are converted into energy-intensive goods and services including operation of vehicles that clean the air.

Renewable energy parks will be located in solar-rich, wind-rich, hydro-rich, wave-rich areas and where abundant biomass wastes are found. Illustratively, nearly ideal combinations of solar and wind resources are found near the Arizona-New Mexico border along Highway I-40. An excellent initial site for Renewable Energy Parks that will facilitate technology transfer demonstrations is in Apache County, Arizona, as shown on the resource maps of Figures A-2a-b and A-3. This location is near natural gas lines and high power electrical transmission lines for wheeling methane and electricity to industrial parks and refueling stations along I-40 and interconnecting corridors. Methane that has been wheeled through the natural gas pipelines will be converted into hydrogen and carbon for sales from refueling stations. Carbon will be utilized for making durable goods. In addition at off-peak times electricity that has been wheeled through

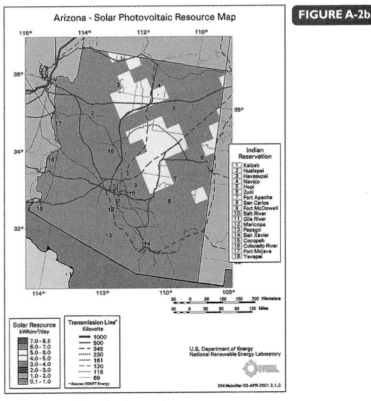

Arizona - Solar Photovoltaic Resource Map

the electricity grid will be utilized to produce hydrogen by electrolysis of water.

213

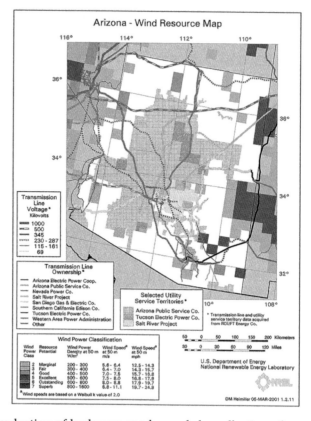

Appendix

FIGURE A-3

Arizona - Wind Resource Map

Primary production of hydrogen can be made by collection of solar and wind energy. Solar dish gensets, photovoltaics, and wind generators will produce electricity. Electronic contracts will determine the destinations of electricity that is wheeled to refueling stations that electrolyze water to produce hydrogen for engine and fuel cell powered vehicles. Hydrogen will be produced by electrolysis at the Renewable Energy Parks to fully utilize production capacity at times that electronic contract deliveries do not demand full capacity. Organic material wastes from nearby communities including garbage, sewage, farm and forest wastes will be processed in solar conversion furnaces into methane that will be transported by natural gas pipelines to industrial parks for production of hydrogen and durable goods that utilize the carbon that is co-produced along with hydrogen. Renewable methane will also be transported to refueling centers for Hy-Boost blends and to co-produce carbon products as hydrogen is released. Equations A-1 and A-2 summarize these overall processes:

214

Organic Waste + Solar Heat $\longrightarrow$ Methane   EQUATION A-1

Methane $\longrightarrow$ Hydrogen + Carbon Products   EQUATION A-2

At times when the Renewable Energy Parks have surplus hydrogen it can be reacted with off-grade coal from nearby mines to produce clean methane. Methane will be piped to industrial parks to release renewable hydrogen and clean carbon for production of durable goods. Equations A-3 and A-4 summarize these clean renewable hydrogen and material production benefits from coal.

Dirty Coal + Hydrogen $\longrightarrow$ Clean Methane   EQUATION A-3

Clean Methane $\longrightarrow$ Hydrogen + Carbon Products   EQUATION A-4

Lighter, stronger, more corrosion-resistant vehicle components will be manufactured along with other carbon products at industrial parks to provide more fuel-efficient, safer, and longer lived, cars, trucks, and aircraft.

Figure A-4 shows the type of modular electrolyzer units that will produce pressurized supplies of hydrogen and oxygen from water and renewable electricity that is wheeled to the refueling stations. After delivery by truck or rail these modular units will be connected to the electric grid and to the water line to rapidly begin refueling operations.

Figure 12.8 summarizes the process provided by modular conversion units that produce carbon products and pressurized supplies of hydrogen from renewable methane that is wheeled to the refueling stations.

These modular systems for quickly enabling hydrogen to be delivered at refueling stations will be mass-produced by manufacturing operations simlar to those now producing 40 million new vehicles each year.

FUTURE CATALOG ITEMS TO BE OFFERED
BY NEW VENTURES THAT IRRI LAUNCHES

Introduction to New Techologies and Machines

In many instances, a Hydrogen Service Stations will be located in a more densely populated area with code restrictions that prohibit large scale primary energy conversion systems.

SOLUTIONS:

Wheeling technologies will provide transport of precursor fuels or electricity to produce hydrogen at such locations.

Renewable methane can readily be produced from sewage, garbage, and agricultural or wastes. Bio-Methane or "biothane" can be added to natural gas lines to deliver renewable hydrogen and carbon. Renewable hydrogen can be also reacted with carbon from a suitable carbon donor such as coal to produce methane. These sources of methane can be added to the national grid of natural gas pipelines where it will be virtually indistinguishable from the fossil methane that is the main constituent of natural gas.

Methane will be delivered by pipeline to the HSS or to a nearby industrial park where it is dissociated into carbon and hydrogen. Hydrogen will be used to re-fuel vehicles at the HSS. Carbon will be converted into a vast variety of durable goods. This provides extremely beneficial carbon sequestration into valuable durable goods along with efficient delivery of renewable hydrogen.

Renewable electricity will be produced at distant locations by solar, wind, wave or biomass conversion systems. One or more pressurizing electrolyzers will be located at the Hydrogen Service Station to produce pressurized supplies of hydrogen.

Hydrogen Service Stations may also be provided with compressed and cryogenic hydrogen deliveries by trucks. In many areas these hydrogen transport trucks will receive hydrogen from bulk plants that are re-fueled by rail, barge, or pipeline transport of hydrogen.

HARDWARE and COMPONENTS

PRESSURIZING ELECTROLYZERS

Commercial capacity carbon fiber reinforced electrolyzers produce pressurized hydrogen and oxygen. Figure A-4 depicts a family of electrolyzers that deliver pressurized supplies of hydrogen and oxygen. Optional sub-systems provide reverse osmosis removal of dissolved substances, pressurization system for feed water, and related pressure piping components.

Pressurized H$_2$

WIND

SOLAR

HYDRO

H_2

O_2

Pressurized O$_2$

Pressurized Electrolyzer

Electric Power Controller

Pump

Water

COMPRESSED GAS (CG) HYDROGEN STORAGE TANKS:

Commercial storage canisters for receiving pressurized supplies of hydrogen are shown in Figure A-5. These carbon fiber reinforced storage tanks come complete with a pressure and/or excessive heat

actuated relief valve, tank shutoff valve and mounting hardware for securing them safely. Unless otherwise specified these tanks are supplied with about two atmospheres of nitrogen to expedite convenient system testing, drain-down, and utilization with hydrogen, methane or other acceptable fuel gases. These tanks may be supplied in any color. The tanks shown are black with permanent identification of serial number, gas storage type, working pressure capacity, and hydrostatic test pressure.

CRYOGENIC HYDROGEN STORAGE TANKS:

Commercial storage canisters for receiving super cold supplies of hydrogen are shown in Figure A-6. These carbon fiber reinforced storage tanks come complete with a pressure and/or excessive heat actuated relief valve, tank shutoff valve and mounting hardware for securing them in a safe place. Unless otherwise specified these tanks are supplied with about two atmospheres of nitrogen to expedite testing and convenient system flush out and utilization with hydrogen, methane or other acceptable cryogenic fuels. The tanks shown are black with permanent identification of serial number, gas storage type, working pressure capacity, and hydrostatic test pressure. Optional sub-systems convert cryogenic liquid supplies to compressed gas to enable a variety of re-fueling operations.

FIGURE A-6

CG OXYGEN STORAGE TANKS

Storage canisters for receiving pressurized supplies of oxygen are shown in Figure A-7. These carbon fiber reinforced storage tanks come complete with a pressure and/or excessive heat actuated relief valve, tank shutoff valve and mounting hardware for securing them in a safe place. Unless otherwise specified these tanks are supplied with about two atmospheres of oxygen to expedite convenient utilization with oxygen.

FIGURE A-7

CRYOGENIC OXYGEN STORAGE TANKS:

Commercial storage canisters for receiving super cold supplies of oxygen are shown in Figure A-8. These carbon fiber reinforced storage tanks come complete with a pressure and/or excessive heat actuated relief valve, tank shutoff valve and mounting hardware for securing them in a safe place. Unless otherwise specified these tanks are supplied with about two atmospheres of oxygen to expedite convenient utilization with cryogenic oxygen. The tanks shown are black with permanent identification of serial number, gas storage type, working pressure capacity, and hydrostatic test pressure. Optional sub-systems convert compressed oxygen to cryogenic liquid to enable a variety of storage and transport operations.

FIGURE A-8

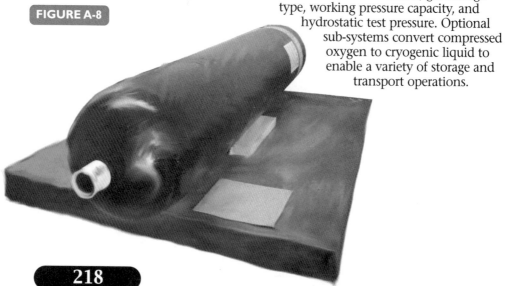

PRESSURE TUBING AND CONVERSION KITS

Seamless, annealed, nonmagnetic stainless steel tubing is recommended along with vibration-resistant, high-temperature polymer tubing with braided reinforcement as shown in Figure A-9 for delivery of hydrogen and/or HyBoost fuels to engines through conversion kits such as shown in Figure A-10.

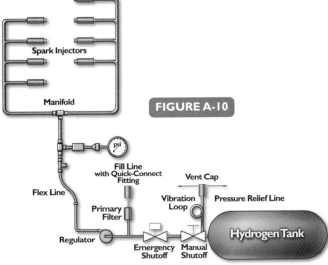

Spark Injectors

Manifold

FIGURE A-10

psi

Fill Line with Quick-Connect Fitting

Vent Cap

Flex Line

Vibration Loop

Pressure Relief Line

Primary Filter

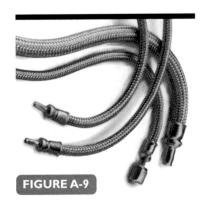

FIGURE A-9

Regulator

Emergency Shutoff

Manual Shutoff

Hydrogen Tank

FUTURE CATALOG OF COMPONENTS FOR FARM, RAIL, TRUCK APPPLICTIONS:

FARM POWER: TOTAL ENERGY SYSTEM FP (TES FP)

This long-life engine-generator produces electricity for your farm. Surplus generating capacity is diverted to a pressurizing electrolyzer that provides high-purity hydrogen needed for many fuel cell applications. TES FP has an integral thermochemical regenerator to convert heat supplied by the engine into Hy-Boost fuel values for providing extremely high overall efficiency. TES FP provides pressurized fuel for quickly and conveniently refueling other farm equipment. TES FP operates efficiently on hydrogen, biomass sources methane, natural gas, propane, butane, or converted biomass feed stocks such as bagasse, field stover, manure, and forest wastes. TES FP provides dehumidified air for drying crops, hot water, and home heating where needed. Optional subsystems include various sizes of hydrogen barbeque and cooking units. Show your TES FP to the neighbors as you cook up your favorite recipes… with clean carcinogen-free hydrogen.

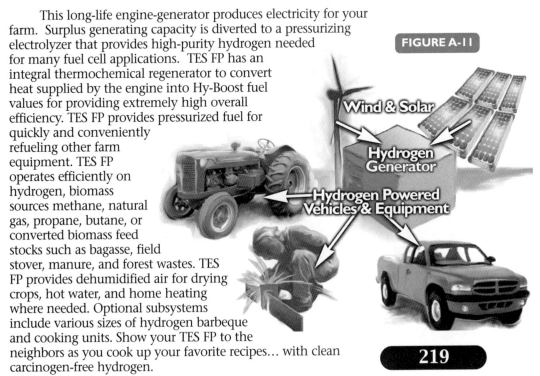

FIGURE A-11

Wind & Solar

Hydrogen Generator

Hydrogen Powered Vehicles & Equipment

219

Appendix

TRACTOR AND IMPLEMENT CONVERSION KITS

Conversions are designed to allow conversion of existing farm engines to operation on farm-produced fuels. Please inquire about specific kits that will soon be available. Figure A-10 depicts the type of kit used for conversion of typical diesel engines.

FIGURE A-12

RAIL ENGINES:

Figure A-13 shows a four-stroke General Electric Locomotive that can be remanufactured for operation on hydrogen, Hy-Boost fuels, methane, natural gas, LP gases, and fuel alcohols. Figure A-14 shows a two-stroke General Motors Locomotive that can be remanufactured for operation on hydrogen, Hy-Boost fuels, methane, natural gas, LP gases and fuel alcohols. These beautiful locomotives will be completely remanufactured and made ready to

FIGURE A-13

haul long train loads of renewable resources to market. In application on U.S. railroads these locomotives can save some $2 billion each year by utilizing renewable fuels made in America by replacing diesel fuel made from imported oil. Locomotives that clean the air will be available for lease or hire. Please note our optional cryogenic or super range compressed fuel storage cars for use as fuel tenders for our new RF series locomotives.

TRUCK TRACTORS:

Remanufactured tractors with hydrogen engines that are more powerful and last longer are shown in Figure A-15. These beautiful hydrogen tractors will be on the road even if diesel fuel is too pollutive or too expensive to burn. Optional items include four sizes of our new super-range hydrogen storage tanks. Lease one today for healthier, happier, more cost effective trucking.

FIGURE A-15

PRODUCTS THAT WILL BE MARKETED BY NEW VENTURES that are launched by the INTERNATIONAL RENEWABLE RESOURCES INSTITUTE

1. Catalog of components for the Solar Hydrogen Home
2. Catalog of components for Hydrogen Service Stations
3. Catalog of components for Hydrogen ICU Emergency Response Systems For Disaster Relief
4. Catalog of piping components and systems for delivering potable water, air-conditioning applications, distributing renewable fuels, collecting wastes, and chemical process applications.
5. Catalog of hydrogen farm, rail, and truck equipment

TYPICAL TECHNOLOGY TRANSFERS

U.S. Patent 5,394,852
*SparkInjector
replacement for spark
plugs, diesel fuel
injectors, and
carbourators allows
existing engines to
utilize low-cost
renewable fuels*

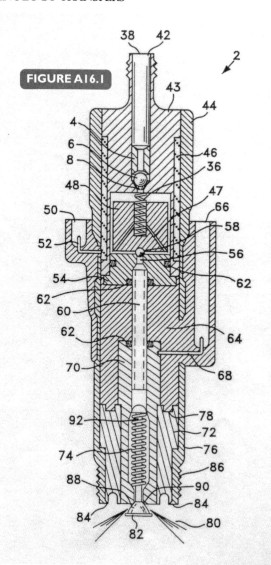

FIGURE A16.1

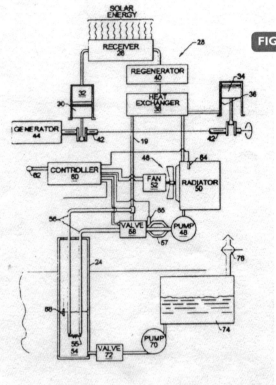

U.S. Patent 5,899,071
*High efficiency solar
energy conversion
system*

U.S. Patent 6,015,065
*System for very
compact storage of
Hydrogen*

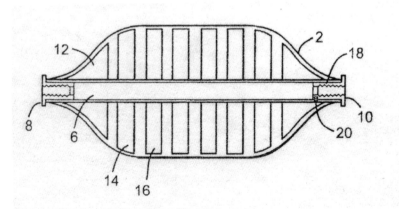

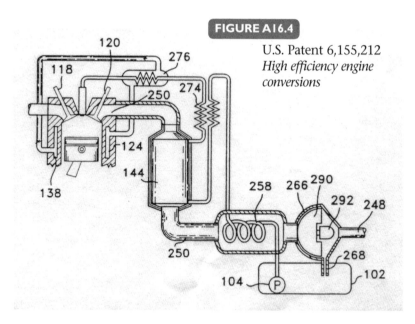

FIGURE A16.4

U.S. Patent 6,155,212
*High efficiency engine
conversions*

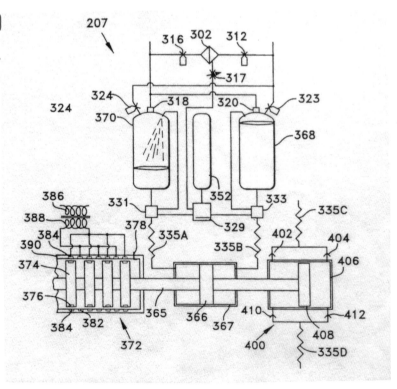

FIGURE A16.5

U.S. Patent 6,446,597
*Advanced energy
conversion sysem*

WIND GENERATORS

Select the wind machines that best meet your needs from the following recommended products. Figure A-17 shows typical carbon fiber reinforced wind generators. Optional subsystems convert wind generator output to household electricity requirements and/or electrolyser of water or waste water.

FIGURE A-17

FALLING WATER GENERATORS

In-stream or directed-water generators for converting the kinetic energy in falling water to electricity are depicted in Figure A-18. Evaluate your stream velocity, depth, wild life protection and debris avoidance requirements, select the components that you need and let nature do the rest. These carbon fiber reinforced systems will last much longer than metal systems.

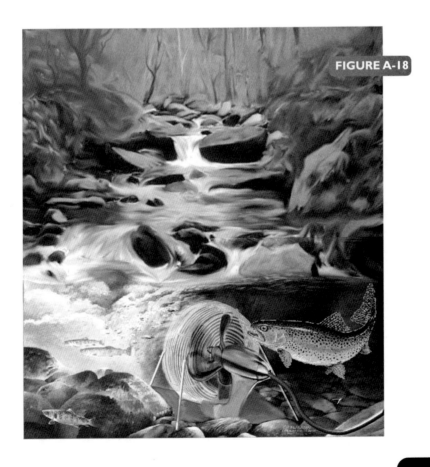

FIGURE A-18

WAVE GENERATORS

Wave generators convert the kinetic energy in wave motion to electricity. Highly durable and efficient carbon fiber reinforced wave generators are depicted in Figure A-19. Evaluate your wave potential, wild life protection, and debris avoidance requirements, select the components that you need and let wind over thousands of square miles power your wave machine.

PRESSURIZING ELECTROLYZERS

Whether you convert solar, wind, wave, or exercise machine energy into electricity in amounts sufficient to meet the peak electrical requirements of your home, most of the time there will be surplus generating capacity that could be utilized to produce oxygen and hydrogen. Figure A-20 depicts a family of carbon fiber reinforced electrolyzers that deliver pressurized supplies of hydrogen and oxygen. Please note the availability of optional sub systems for reverse osmosis removal of dissolved substances, pressure pumps, and related pressure piping components.

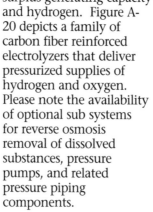
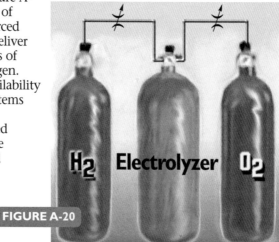

H_2 Electrolyzer O_2

FIGURE A-20

COMPRESSED STORAGE:

A storage canister for receiving pressurized supplies of hydrogen is shown in Figure A-21. This carbon fiber reinforced storage tank comes complete with a pressure and/or excessive heat actuated relief valve, tank shutoff valve and mounting hardware for securing it in a safe place. Unless otherwise specified such tanks are supplied with about two atmospheres of nitrogen to expedite convenient system testing, calibration and utilization with hydrogen, methane or other acceptable fuel gases. Unless otherwise specified these tanks are black with permanent identification of serial number, gas storage type, working pressure capacity, and hydrostatic test pressure. Please note the availability of optional pressure tubing, fittings, tube cutters, tube benders, annealing torches, pressure gages, electronic pressure transducers, pressure regulators and solenoid operated control valves.

FIGURE A-21

FIGURE A-22

COMPRESSED OXYGEN STORAGE

Storage canisters for receiving pressurized supplies of oxygen are shown in Figure A-22. These carbon fiber reinforced storage tanks come complete with a pressure and/or excessive heat actuated relief valve, tank shutoff valve and mounting hardware for securing them in a safe place. Unless otherwise specified these tanks are supplied with about two atmospheres of nitrogen to expedite convenient system testing and utilization with oxygen. Please note the availability of optional pressure tubing, fittings, tube cutters, tube benders, annealing torches, pressure gages, electronic pressure transducers, pressure regulators and solenoid operated control valves.

HYDROGEN BARBEQUE AND COOKING CENTER

Hydrogen provides revolutionary cooking improvements. Combustion of hydrogen produces steam. Steam can be utilized to moisturize and retain the natural flavors of food being cooked. Treat yourself to better health by using carcinogen-free cooking fuel that flavorizes your food.

FIGURE A-23

TOTAL ENERGY SYSTEM INNOVATION (TESI)

TESI's heart is a long-life engine and generator. TESI's engine utilizes carbon/graphite pistons and cylinders for extremely long life and efficient operation. The exhaust of this hydrogen-burning engine is used to heat your cooking center, bath water, sauna, or home. Electricity produced by TESI can match your peak needs or all of your needs. TESI can use renewable fuels such as hydrogen, methane, or fossil fuels such as natural gas and propane. TESI can double the energy utilization efficiency compared to conventional central power plants.

Air Filter & Conditione

Clean Air (ou

Radiant Heating from hot water circuits in floor

Hydrogen Electricity Generator & Heat Exchanger

Polluted Air (in)

Spa & Hea

TOTAL ENERGY FUELCELL SYSTEM (TEFS)

TEFS utilizes an advanced fuel cell with extensive carbon/graphite componentry advantages to generate electricity. TEFS can also operate in reverse mode to convert water into hydrogen and oxygen. Heat left over from the electricity production operations is used to heat your cooking center, bath water, sauna, or home. Electricity produced by TEFS can match your peak needs or all of your needs. TEFS can use renewable fuels such as hydrogen, methane, or fossil fuels such as natural gas and propane. TEFS can double the energy utilization efficiency compared to conventional central power plants.

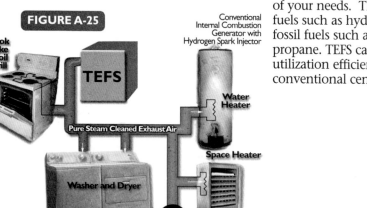

FIGURE A-25

Cook Bake Broil Grill

TEFS

Conventional Internal Combustion Generator with Hydrogen Spark Injector

Water Heater

Pure Steam Cleaned Exhaust Air

Space Heater

Washer and Dryer

Distilled Water Condenser

GET IN SHAPE AS YOU CLEAN THE AIR

EXERCISE GENERATORS

Figure A-26, A-27 and A-28 show carbon fiber reinforced exercise machines that convert human conditioning energy into electricity that is utilized to electrolyze water to produce hydrogen for household applications and oxygen for numerous applications. Continuous monitoring of vital functions are provided to assist in physical conditioning of the human dynamo and compared with useful energy stored as hydrogen.

FIGURE A-26

FIGURE A-27

Get in shape, stay in shape, and produce your own solar hydrogen.

FIGURE A-28

229

COMPONENTS FOR HYDROGEN RE-FUELING DEPOTS AND SERVICE STATIONS IN THE SOLAR HYDROGEN CORRIDOR

PRIMARY COLLECTION OF ENERGY:

Collection of energy can be at or near a Hydrogen Station or at one or more distant places. The following technologies could be utilized if there is proper zoning along with sufficient space and renewable energy for the purpose. If you choose to collect energy at or near your Hydrogen Station consider the following commercial size collection options:

Photovoltaic Systems

Wind Generators

Falling Water Generators

Wave Machines

Biomass Converters

If you choose to utilize energy that is collected elsewhere by one or more of the above listed technologies, arrangements can be made to transport the energy to you by one or more of the following transport options.

Underground transport of renewable methane by natural gas pipeline

Electric grid transport of renewable electricity

Barge, truck, or rail delivery of hydrogen or methane

Electronic Contact Meters

Whether you are billing or being billed for more cost-effective, renewable, anti-inflationary energy from your farm, home, business or exercise machine we have the right electronic contract meter option for you. Please review the options depicted and further explaned at our website - www.americanhydrogenassociation.org.

ACTUAL COST OF GASOLINE AND DIESEL FUEL:

When America was energy independent and the world's largest oil producer and exporter, the "taking cost" and not the "replacement cost" was established as the accepted "cost basis" and the price at the pump ignored numerous actual costs to taxpayers such as the oil-depletion allowance, government grants, sweetheart oil leases, military expenditures to protect or exert U.S. oil interests, and health care expenses for illnesses and injuries caused by oil production and delivery along with air and water pollution from production and use of petrol fuels.

Numerous sources including studies by Lester R. Brown, president of Earth Policy Institute[1], the International Center for Technology Assessment[2], the American Lung Association[3] the Union of Concerned Scientists[4] the American Solar Energy Society[5] and the International Association for Hydrogen Energy[6] have determined that the price paid at the pump for gasoline is far less than the actual cost of petrol fuels. Such sources provide credible accounts of some $559 billion to $1.69 trillion per year for:

Tax Subsidies Given To The Oil Industry

Government Grants and Programs That Assist The Oil Industry

Military and Related Foreign Aid To Protect and Facilitate Importation Of Oil

Health and Environmental Costs

Corrosion and Degradation of Capital Goods Due to Fossil Fuel Pollution

Therefore on annual U.S. sales of about 143 billion gallons of petrol fuels, the hidden price plus the pump price amounts to $5.60 to more than $15 per gallon depending upon how extensively hidden costs are accounted.[6]

How did we become so burdened with hidden costs and subsidies to conceal the true cost of petrol fuels? "The money power of the country will endeavor to prolong its rule by preying upon the prejudices of the people until all wealth is concentrated in a few hands and the Republic will be destroyed."[7] Mark Twain summarized this problem as it evolved by noting "I think I can say, and say with pride, that we have legislators that bring higher prices than any in the world."[8]

REFERENCES:

1. Lester R. Brown, Remarks to the Renewable Hydrogen Forum (2003) based on Lester R. Brown, Plan B: Rescuing a Planet under Stress and a Civilization in Trouble (W.W. Norton & Company, NY: 2003) [forthcoming September 2003].

2. International Center for Technology Assessment Report No. 3 "An analysis of the Hidden External Costs Consumers Pay To Fuel Their Automobiles" (2001)

3. Dr. Anthony DeLucia, American Lung Association Report to the Renewable Hydrogen Forum (2003)

4. Roland Hwang, "Money Down The Pipeline," Union of Concerned Scientists (1995) p. ES-1

5. Michael Nicklas, American Solar Energy Society Report to the Renewable Hydrogen Forum (2003)

6. T. Nejat Veziroglu, International Association for Hydrogen Energy

7. President Abraham Lincoln, November 21, 1864

8. Mark Twain

Appendix

Properties of Fuels

Property	Gasoline	No. 2 Diesel Fuel	Methanol	Ethanol	MTBE	Propane	Compressed Natural Gas (CNG)	Hydrogen
Chemical Formula	C_4 to C_{12}	C_3 to C_{25}	$CH_3 OH$	C_2H_5OH	$(CH_3)_3COCH_3$	C_3H_8	CH_4	H_2
Molecular Weight	100–105(a)	≈200	32.04	46.07	88.15	44.1	16.04	2.02(x)
Composition, Weight %								
Carbon	85–88(b)	84–87	37.5	52.2	66.1	82	75	0
Hydrogen	12–15(b)	33–16	12.6	13.1	13.7	18	25	100
Oxygen	0	0	49.9	34.7	18.2	–	–	0
Specific gravity, 60°F/60°F	0.72–0.78(b)	0.81–0.89(d)	0.796(c)	0.796(c)	0.744(m)	0.508	0.424	0.07(u)
Density, lb/gal @60°F	6.0–6.5(b)	6.7–7.4(d)	6.63(b)	6.61(b)	6.19(m)	4.22	1.07(r)	–
Boiling temperature, °F	80–437(b)	370–650(d)	149(c)	172(c)	131(c)	-44	-259	-423(u)
Reid vapor pressure, psi	8–15(k)	0.2	4.6(o)	2.3(o)	7.8(e)	208	2,400	
Octane no. (1)								
Research octane no.	90–100(u)	–	107	108	116(t)	112	–	130+
Motor octane no.	81–90(s)		92	92	101(t)	97	–	–
(R +M)/2	86–94(s)	N/A	100	100	108(t)	104	120+	
Cetane no. (1)	5–20	40–55	–	–	–	–	–	–
Water solubility, @70°F								
Fuel in water, vol. %	Negligible	Negligible	100(c)	100(b)	4.3(e)	–	–	–
Water in fuel, vol. %	Negligible	Negligible	100(c)	100(b)	1.4(e)	–	–	–
Freezing point, °F	-40(g)	-40–30(4)	-143.5	-173.2	-164(c)	-305.8	-296	-435(v)
Viscosity								
Centipoise @60°F	0.37–0.44(3,p)	2.6–4.1	0.59(j)	1.19(j)	0.35(j)	–	–	–
Flash point, closed cup, °F	-45(b)	165(d)	52(o)	55(o)	-14(e)	-100 to -150	-300	–
Autoignition temperature, °F	495(b)	≈600	867(b)	793(b)	815(e)	850–950	1,004	1,050–1,080(u)
Flammability limits, volume %								
Lower	1.4(b)	1	7.3(o)	4.3(o)	1.6(e,k)	2.2	5.3	4.1(u)
Higher	7.6(b)	6	36(o)	19(o)	8.4(e,k)	9.5	15	74(u)
Latent heat of vaporization								
Btu/gal @60°F	≈900(b)	≈700	3,340(b)	2,378(b)	863(5)	775	–	–
Btu/lb @60°F	≈150(b)	≈100	506(b)	396(b)	138(5)	193.1	219	192.1(v)
Btu/lb air for stoichiometric mixture @60°F	≈10(b)	≈8	78.4(b)	44(b)	11.8	–	–	–
Heating value (2)								
Higher (liquid fuel-liquid water) Btu/ lb	18,800–20,400	19,200–20000	9,750 (2)	12,800 (q)	18,290 (h)	21,600	23,600	61,002 (v)
Lower (liquid fuel-water vapor) Btu/ lb	18,000–19,000	18,000–19,000	8,570 (b)	11,500 (q)	15,100 (h)	19,800	21,300	51,532 (v)
Higher (liquid fuel-liquid water)Btu/ gal	124,800	138,700	64,250	84,100	–	91,300		
Lower (liquid fuel-water vapor) Btu/ gal @ 60° F	115,000	128,400	56,800 (3)	76,000 (3)	93,500 (4)	84,500	19,800 (6) –	
Heating value, stoichiometric mixture								
Mixture in vapor state, Btu/ cubic foot @ 68° F	95.2 (b)	96.9 (5, q)	92.5 (b)	92.9 (b)	–	–	–	–
Fuel in liquid state, Btu/ lb or air	1,290 (b)	–	1,330 (b)	1,280 (b)	–	–	–	–
Specific heat, Btu/ lb °F	0.48 (g)	0.43	0.6 (j)	0.57 (j)	0.5 (j)	--	--	--
Stoichiometric air/ fuel, weight	14.7 (3)	14.7	6.45 (l)	9 (l)	11.7 (j)	15.7	17.2	34.3 (u)
Volume % fuel in vaporized stoichiometric mixture	2 (b)	–	12.3 (b)	6.5 (b)	2.7 (j)	–	–	–

NOTES:

(1) Octane values are for pure components. Laboratory engine Research and Motor octane rating procedures are not suitable for use with neat oxygenates. Octane values

obtained by these methods are not useful in determining knock- limited compression ratios for vehicles operating on neat oxygenates and do not represent octane performance of oxygenates when blended with hydrocarbons. Similar problems exist for cetane rating procedures.

(2) The higher heating value is cited for completeness only. Since no vehicles in use, or currently being developed for future use, have powerplants capable of condensing the moisture of combustion, the lower heating value should be used for practical comparisons between fuels.

(3) Calculated.

(4) Pour Point, ASTM D 97 from Reference (c).

(5) Based on cetane.

(6) For compressed gas at 2,400 psi.

SOURCES:

(a) The basis of this table and associated references was taken from: American Petroleum Institute (API), Alcohols and Ethers, Publication No. 4261, 2nd ed. (Washington, DC, July 1988), Table B- 1.

(b) "Alcohols: A Technical Assessment of Their Application as Motor Fuels," API Publication No. 4261, July 1976.

(c) Handbook of Chemistry and Physics, 62nd Edition, 1981, The Chemical Rubber Company Press, Inc.

(d) "Diesel Fuel Oils, 1987," Petroleum Product Surveys, National Institute for Petroleum and Energy Research, October 1987.

(e) ARCO Chemical Company, 1987.

(f) "MTBE, Evaluation as a High Octane Blending Component for Unleaded Gasoline," Johnson, R. T., Taniguchi, B. Y., Symposium on Octane in the 1980's, American Chemical Society, Miami Beach Meeting, September 10- 15, 1979.

(g) "Status of Alcohol Fuels Utilization Technology for Highway Transportation: A 1981 Perspective," Vol. 1, Spark- Ignition Engine, May 1982, DOE/ CE- 56051- 7.

(h) American Petroleum Institute Research Project 44, NBS C- 461.

(i) Lang's Handbook of Chemistry, 13th Edition, McGraw- Hill Book Company, new York, 1985.

(j) "Data Compilation Tables of Properties of Pure Compounds," Design Institute for Physical Property Data, American Institute of Chemical Engineers, New York, 1984.

(k) Petroleum Product Surveys, Motor Gasoline, Summer 1986, Winter 1986/ 1987, National Institute for Petroleum and Energy Research.

(l) Based on isocetane.

(m) API Monograph Series, Publication 723, "Teri- Butyl Methyl Ether," 1984.

(n) BP America, Sohio Oil Broadway Laboratory.

(o) API Technical Data Book – Petroleum Refining, Volume I, Chapter I. Revised Chapter 1 to First, Second, Third and Fourth Editions, 1988.

(p) "Automotive Gasolines," SAE Recommended Practice, J312 May 1986, 1988 SAE Handbook, Volume 3.

(q) "Internal Combustion Engines and Air Pollution," Obert, E. F., 3rd Edition, Intext Educational Publishers, 1973.

(r) Value at 80 degrees F with respect to the water at 60 degrees F (Mueller & Associates).

(s) National Institute for Petroleum and Energy Research, Petroleum Product Surveys, Motor Gasolines, Summer 1992, NIPER- 178 PPS 93/ 1 (Batlesville, OK, January 1993), Table 1.

(t) P. Dorn, A. M. Mourao, and S. Herbstman, "The Properties and Performance of Modern Automotive Fuels," Society of Automotive Engineers (SAE), Publication No. 861178

(Warrendale, PA, 1986), p. 53.

(u) C. Borusbay and T. Nejat Veziroglu, "Hydrogen as a Fuel for Spark Ignition Engines," Alternative Energy Sources VIII, Volume 2, Research and Development (New York: Hemisphere

Publishing Corporation, 1989), pp. 559- 560.

(v) Technical Data Book, Prepared by Gulf Research and Development Company, Pittsburgh, PA, 1962.

Appendix

SELECTED ENERGY EQUIVALENTS

Compared Energy Source	Cubic Meter H₂ Gas	Cubic Foot H₂ Gas	Liter Liquid H₂	Gallon Liquid H₂	Kilogram H₂	Pound H₂
Gasoline Liters	0.352	0.00929	0.279	1.06	3.93	1.78
Methanol Liters	0.676	0.0178	0.536	2.03	7.55	3.41
Diesel Liters	0.279	0.00737	0.221	0.837	3.12	1.41
Jet Fuel Liters	0.287	0.00757	0.227	0.860	3.20	1.45
Methane (scf)	11.4	0.301	9.05	34.2	128	57.6
Propane (scf)	4.48	0.118	3.55	13.4	50.1	22.6
Butane (scf)	3.45	0.091	2.73	10.3	38.5	17.4
Coal Anthracite (Tons)	0.000397	0.0000105	0.000315	0.00119	0.00444	0.0020
Coal Bituminous (Tons)	0.000392	0.0000104	0.000311	0.00118	0.00438	0.00198
Coal Lignite (Tons)	0.000731	0.0000193	0.000579	0.00219	0.00816	0.00369
Barrels of Crude	0.00176	0.0000466	0.00140	0.00529	0.0197	0.00890
Gasoline Gallons	0.0930	0.00246	0.0737	0.279	1.04	0.469
Methanol Gallons	0.179	0.00471	0.142	0.535	1.99	0.901
Diesel Gallons	0.0738	0.00195	0.0584	0.221	0.824	0.372
Jet Fuel Gallons	0.076	0.00200	0.0600	0.227	0.846	0.382
H2 Gas Cubic Meters (STP)	1.0	0.0264	0.792	3.0	11.2	5.04
H2 Gas Cubic Feet (NTP)	37.9	1.0	30.0	114	423	191
H2 Liquid Liters (nbp)	1.26	0.0333	1.0	3.78	14.1	6.40
H2 Liquid Gallons (nbp)	0.334	0.00880	0.264	1.0	3.72	1.69
H2 Kilograms	0.0896	0.00236	0.0709	0.268	1.0	0.454
H2 Pounds	0.198	0.00521	0.156	0.592	2.20	1.0
H2 Tons	0.0000987	0.0000026	0.0000782	0.000296	0.0011	0.00050
Electricity KW-hours	3.00	0.0791	2.38	8.99	33.5	15.1
Electricity MW-hours	0.003	0.0000791	0.00238	0.00899	0.0335	0.0151
H2 High HV gigajoules	0.0128	0.00034	0.0101	0.0383	0.143	0.0644
H2 High HV million Btus	0.0121	0.000319	0.0096	0.0363	0.135	0.0610
H2 High HV Btu	12,100	319	9,600	36,300	135,000	61,000
H2 High HV kilocalories	3,100	80.5	2,400	9,100	34,100	15,400
H2 Low HV gigajoules	0.0108	0.000285	0.0086	0.0324	0.121	0.0544
H2 Low HV million Btus	0.0102	0.000270	0.0081	0.0307	0.114	0.0516
H2 Low HV Btu	10,200	270	8,100	30,700	114,000	51,600
H2 Low HV kilocalories	2,600	68	2,040	7,700	28,800	13,000

References 1. H-Ion Solar Incorporated - 6095 Monterey Ave., Richmond, CA 94805, USA 2. The Hydrogen World View by Roger Billings - American Academy of Science 1991 3. Diesel Fuels Technical Review (FTR-2) by Chevron Products Company a Division of Chevron USA Inc. 1998 4. Motor Gasoline Technical Review (FTR-1) by Chevron Products Company a Division of Chevron USA Inc. 1998

To use this comparison at standard temperature and pressure, substitute the numeric value in the table to quantify the energy source listed on the left. The result is the equivalent hydrogen fuel, in units shown at the top.

For example, to get the same energy as one pound of hydrogen, substitute the equivalent, .469 gallons of gasoline, or 57.6 scf methane.

Data sources from :

The Hydrogen World View by Roger Billings - American Academy of Science 1991

Diesel Fuels Technical Review (FTR-2) by Chevron Products Company a division of Chevron USA Inc 1998

Motor Gasolines Technical Review (FTR-1) by Chevron Products Company a division of Chevron USA Inc 1996

THE BIGGEST STORY IN THE WORLD.

My name is Ray Smucker. I entered the broadcast industry as a radio entertainer with the Horace Heidt novelty show doing voice imitations of celebrities, politicians and other radio voices.

In the summer of 1942 my wife and I were traveling from Iowa to California and we stopped in Dallas, Texas to visit Charles Dabney and his wife. We discussed the terrible war raging in the Pacific islands with Japan and in Europe with Germany and how the world was at war over oil.

Mr. Dabney asked if I wanted to see something interesting and soon we were in an inventor's garage where three Mercury engines were being tested. The inventor took an empty lye can and nearly filled it with water. He carefully poured the water into a glass box that replaced the carburetor. This box had numerous plates in it and it started to fizz. He used a piece of paper to cover the box and struck a match. When the match approached the edge of the paper it was blown into the air. Next he fitted a cover on the box and started the engine. It ran smoothly and the oil was clean.

As we left Dallas I told my wife, "When they make the hydrogen from sunshine and water it will be the biggest story in the world."

The next time I talked to Charles Dabney he said that the inventor had his picture in the paper with a truck that ran on hydrogen, and 22 men were riding on the truck bed. The paper reported that the inventor was going to Detroit to demonstrate his invention. About two weeks later he said he went to the inventor's shop and it was empty and the inventor was gone. It was rumored that he received quite a lot of money.

I became a news and information broadcaster and for 50 years I watched for information about hydrogen. One day in 1993 I saw an article in the Phoenix paper about the American Hydrogen Association and I called their office. I told Roy McAlister what I had been thinking about for 50 years and he said, "Come on out and I will show you how the Dallas inventor did it and how to make hydrogen from solar energy."

I went to the American Hydrogen Association's headquarters and Roy McAlister showed me how to use a large solar concentrator to make electricity that he used in an electrolyzer to make hydrogen from water. He showed me how the hydrogen was stored in red pressure cylinders. Some of these cylinders were very old. They were made in 1916, 1917, and 1919. Roy added hydrogen from one of the cylinders to a tank on a Dodge pickup, and he asked if I wanted to drive the truck. I did, and it ran well. After we stopped, I kept the

Editor's note: Ray Smucker has been an active participant in the Arizona business community for over half a century. For many years Ray served as manager of the key Television station in the Phoenix area. He has been active in the Rotary organization, and a multitude of Civic Organizations. He has been a 'promoter' of hydrogen over all these years. Ray's enthusiasm and participation has been an inspiration for the American Hydrogen Association.

Perspective

engine running, breathed the exhaust, collected water that condensed in the tail pipe, and I drank the water. I went home and told my wife, "This is the biggest story in the world!"

During the last ten years I have learned much more about hydrogen and what is possible from the American Hydrogen Association. The truth is that right now we could be running our cars on hydrogen that cleans the air and makes good water. Sadly, however, the land of the free doesn't know about this truth and we continue to believe that we have no choice but to burn fossil fuels.

The all-powerful fossil fuel industry has kept the public from having hydrogen, and now we are at war and will be at war until we decide to replace oil with hydrogen.

Everyone involved can stop killing people for oil and get busy making a far better world that runs on hydrogen. Roy McAlister tells how to do it in this book. It is the biggest story in the world.

Ray Smucker,

The Pollution-Free Planet Foundation, *founded in*

1995 as the Pollution-Free Planet Committee of the Phoenix Rotary Club, promotes the use of Renewable Energy, especially Hydrogen, to replace fossil fuels around the world.

The inspiration for our organization came from a presentation by Roy McAlister at one of our Club meetings in the spring of 1995. We all had the same questions and misgivings that we were to see in countless others over the years. Six months later, through the perseverance of Ray Smucker and Craig Wilson, the first meeting of the Pollution-Free Planet Committee was held.

Our original mission was to educate Rotarians around the world about the many advantages of Hydrogen. Since then, with the help of Roy and the American Hydrogen Association, we have spoken to approximately 80 Rotary Clubs and a dozen other organizations. We have hosted information booths at six Rotary District Conferences, a Zone Conference, and two International Conventions.

The Pollution-Free Planet Foundation greatly appreciates the invaluable help we have received from Roy McAlister and the American Hydrogen Association. We could not have reached nearly the people we have without Roy's help and we anxiously await the publication of his new book.

William M. Chase Jr. *President,*
Pollution-Free Planet Foundation

William M. Chase Jr., P.E. is Vice President and Chief Electrical Engineer at Peterson Associates Consulting Engineers in Phoenix, Arizona, U.S.A. He has 40 years experience in industrial and construction electrical engineering. He is a founding member of the Pollution-Free Planet Committee and was Chair for six of its seven years. He currently serves as President of the Pollution-Free Planet Foundation.

From Bryan Beaulieu*

" Two years ago I set out to find the best systems and materials from which to build a home that would be the healthiest for my family and our planet. Fortunately I met Roy McAlister. He is the Thomas Edison and Wright Brothers of the new age of hydrogen. If half his inventions were to make it into the market place our world would be transformed.

My home will become a test bed for Roy's incredible concepts and machines. The hydrogen house, as it has become known, will not only create a luxurious, pollution-free living environment, it will also clean the air and water that pass through it. Through living in a "minus emissions" hydrogen fueled home, my family can help Earth become a little cleaner.

Roy has dedicated his life to showing his fellow human beings how we can live in peace with abundance for all. I hope we have time to make his ideas work."

*Bryan Beaulieu is a very successful businessman, inventor and an engineer with an inquiring mind. Recently he sold his Minnesota based business and moved to Arizona. Like all inventions, which are born out of perceived need, Bryan's pollution free, and energy self-sufficient house, owes its conception to a need. Bryan's wife suffers from acute allergies. His quest for the most environmentally friendly energy sources and design led him to Roy McAlister, and the use of hydrogen. His new home will be an example to all, as to "what is possible" in our new hydrogen economy. He is not only fueling his home with Solar & Biomass produced hydrogen, he is converting his automobiles to operate on hydrogen, produced primarily at his home.

Components for your Solar Hydrogen Home

PRIMARY COLLECTION
OF ENERGY:

Collection of energy can be at your home or a different location. If you choose to collect energy at your home consider the following collection options.

• Solar, Lighting, Heating and Cooling • Photovoltaic Systems
• Wind Generators • Falling Water Generators • Wave Machines
• Biomass Converters • Exercise Machines

INSTITUTE OF ECOLONOMICS:

The Economics of Sustainability

The Institute of Ecolonomics provides studies in energy conservation, education, and the advancement of alternative fuel technologies. This spring we met with Spencer Abraham, U.S. Secretary of Energy, at the completion of our "Drive to Survive" demonstrations of alternative fueled vehicles that traveled from Los Angeles to Washington, D.C. We discussed the urgent need to develop a sustainable economy by achieving worldwide commitment to production, distribution, and utilization of renewable fuels.

Conversion to The Solar Hydrogen Civilization that is advocated in this book is not only possible but essential. This conversion can be accomplished in less than two decades. It is imperative for freedom loving persons to become informed and committed to a Grand Plan to develop sustainable worldwide prosperity. The health and welfare of children throughout the world depends upon implementing this very real plan for transition to renewable hydrogen.

Dennis Weaver*

Chairman, CEO

Institute of Ecolonomics

*EDITORS NOTE: *Most of us will recognize Dennis Weaver as both a movie and television star. Lesser known about Dennis Weaver is his 40 year dedication to "making a better planet." Dennis has tirelessly worked through his Foundation towards world-wide economic sustainability, and cleaner, sustainable fuels. He has been a long-time supporter of Roy McAlister's work.*

Perspective

EARTH POLICY:

Environmentally Sustainable Economy

"Earth Policy Institute, an independent environmental research organization founded in May 2001 by Lester R. Brown (founder and former president of Worldwatch Institute), is dedicated to building an environmentally sustainable economy—an eco-economy."

Lester Brown says of this book:

"Current environmental research deals with various pieces of what it takes to build a sustainable future, but it does not provide a model or an overall plan. Unless we have a shared vision of where we want to go, we are not likely to get there. Earth Policy Institute is dedicated to providing this vision via its books, including its flagship publication Eco-Economy: Building an Economy for the Earth, and through its brief Eco-Economy Updates and Eco-Economy Indicators. This information is available for free downloading at http://www.earth-policy.org.

EPI's interdisciplinary research program puts it on the cutting-edge of environmental research. Its researchers can often see new trends emerging well before those who engage in more specialized research. Published in over 18 languages, its publications are reprinted in newspapers, magazines, and websites worldwide, are read by decisionmakers worldwide, are used in college courses, and are relied on by the international media.

Lester R. Brown
President,
Earth Policy Institute
May 2003

One of the keys to an eco-economy is shifting from the current fossil fuel-based economy to a hydrogen economy. The Solar Hydrogen Civilization brings us a step closer to implementing this shift. This new book, just released by the American Hydrogen Association, shows that we can now produce enough hydrogen from renewable resources to replace our current dependence on fossil fuels. It also demonstrates that existing internal combustion engines can be easily converted to operation on hydrogen. This conversion should now begin even as we wait for the mass production of fuel cell vehicles."

Lester R. Brown, "Building a Sustainable Society" ISBN 0-393-01482-7, W.W. Norton & Company 1981

Index

Want to know more about the coming Hydrogen Revolution?

See us on the Web at
www.clean-air.org
www.GoH2.org
www.americanhydrogenassociation.org

Or write to us at:
American Hydrogen Association
P.O. Box 41896
Mesa, AZ 85274